Concrete
A Homeowner's
Illustrated Guide

Concrete
A Homeowner's Illustrated Guide
Deluxe Edition

David H. Jacobs Jr.

TAB **TAB BOOKS**
Blue Ridge Summit, PA

FIRST EDITION
FIRST PRINTING

© 1992 by **TAB Books**.
TAB Books is a division of McGraw-Hill, Inc.

Library of Congress Cataloging-in-Publication Data

Jacobs, David H.
    Concrete : a homeowner's illustrated guide / by David H. Jacobs,
  Jr.—Deluxe ed.
      p. cm.
    Includes index.
    ISBN 0-8306-3910-1 (hard)
    1. Concrete construction—Amateurs' manuals. I. Title.
TA682.42.J32   1992
693'.5—dc20   91-38061
                                         CIP

TAB Books offers software for sale. For information and a catalog, please contact TAB Software Department, Blue Ridge Summit, PA 17294-0850.

Acquisitions Editor: Kimberly Tabor
Book Editor: Lori Flaherty
Production: Katherine G. Brown
Book Design: Jaclyn J. Boone
Cover Photography: Susan Riley, Harrisonburg, VA
Cover Design: Cindy Staub, Hanover, PA   TAB1

Contents

Acknowledgments

Seldom can a concrete job of any size be accomplished without help. The same holds true for writing a book about concrete work. Therefore, I would like to thank a few special people for their assistance and support throughout this project.

Keith Rieber and Gerald Brewster took pictures of some of the concrete jobs featured in this book. I appreciate their time and effort, as it is not always easy to stop in the middle of projects to photograph progress.

Gratitude is also extended to Dave Reum and his crew from Custom Concrete for being so patient while I photographed them during concrete forming, pouring, and finishing operations. The same goes for the Gamby brothers.

The people at Southern Oregon Concrete were equally helpful, spending time with me to explain their operations and how the concrete business is viewed from their perspective.

Thanks to Van and Kim Nordquist from Photographic Designs for doing an excellent job processing the film and printing pictures, taking extra time to make sure each photo turned out just right.

Janna Jacobs provided lots of needed support and plenty of help with the overall organization and photo coordination.

I must also thank Kim Tabor from TAB Books for her support and editorial assistance. Her enthusiasm and professional insight helped to make this an enjoyable project.

Introduction

*I*n the last few years, concrete has become so expensive that some have referred to it as *gray gold*. Because of the cost and professional labor fees, many homeowners are going without concrete walkways and patios. If your concrete job would cost only half as much as a professional's estimate, maybe you could afford to get it done right away.

Most homeowner's concrete needs are about the same—a walkway down the side of a house, a driveway, or a patio in the backyard. These are rather simple jobs that conscientious do-it-yourself homeowners can easily do.

Far too often, homeowners have expressed that they would like to do concrete flat work jobs themselves, but ... Their "buts" are, for the most part, simple problems with easy solutions. Almost 50 percent of the cost of most concrete flat work projects goes toward professional labor fees, and because form lumber and stakes can be used for future projects to offset their initial expense, do-it-yourself concrete finishers can realistically expect to save up to 50 percent off professionally installed concrete.

Concrete flat work includes projects such as, patios, walkways, driveways, and mowing strips. Books about masonry sometimes include a chapter or two on concrete flat work, but they generally don't go into great detail about flat work designs, concrete yardage calculations, order delivery, pouring, finishing, and other basic concrete flat work.

From page 1, this book takes you through each and every step you'll need to form, pour, and finish quality concrete flat work. From designing jobs to final cleanup, everything is described, explained, and illustrated. You'll learn how to design slabs and walkways that conform with your yard's landscape, how to properly form creative designs, determine how much concrete is needed, how to estimate overall costs, determine the amount of help you'll need, how to pour and finish concrete flat work, how to protect it, and how to add some custom touches.

You might not qualify as a concrete professional after reading this book—you simply can't learn all there is to know about a professional trade by reading one book—but by following directions and thinking jobs through, you'll be able to complete professional and functional concrete flat work.

Should you need help, information is included on what type of help you can expect to find in your area. In many cases, a simple phone call to the right person can get your questions answered and problems solved quickly. If you are not sure about how a particular phase of your concrete project is supposed to be completed, don't be afraid to ask for help or advice. Professional concrete dispatchers are generally quite knowledge-able about their trade and seldom fail to offer useful advice to customers with special problems. If need be, especially for truly unique flat work projects, ask a professional concrete finisher to survey your job site and offer solutions to puzzling questions. Their fee for this service should be quite minimal and well worth it in the long run.

This book was written for you, the do-it-yourself homeowner. Follow the instructions, advice, and recommendations to form, pour, and finish your own concrete work and you're sure to get the self-satisfaction of completing projects on your own, as well as saving money.

Chapter 1

Design considerations

You must allot ample time for planning the type of concrete job ultimately desired. So, get a chair, pencil, paper, and something cold to drink. Then, sit near the proposed work site and start drawing plans (FIG. 1-1). They do not have to be perfect or to scale—just usable and easy to read. Sketch a number of various designs. Add or delete planters, experiment with different widths and lengths, and consider what the overall job will look like with the existing or proposed landscape. Keep in mind that innovative concrete flat work designs can go a long way toward improving the general appearance of almost any landscape while still providing usable paths, entertainment areas, or work spaces.

WALKWAYS

Although basically simple in design, walkways can be attractively designed to offer more than just a suitable means to get from the backyard to the front yard. For example, leaving room for a planter next to your house or a fence would be more attractive than butting it up against the foundation (FIG. 1-2). Leaving open spaces for flowers and shrubs breaks up solid concrete with greenery and color. There are several factors that must be considered first, however.

Will a particular side of the house get enough sunshine to support plants? Will visitors see and use the walkway? Will it simply be a space to store garden tools and trash cans? For the latter, a wide walkway extending from the foundation outward might be best to maximize storage space and still allow for plenty of yard space (FIG. 1-3.).

Width

Intended walkway use helps determine its width (FIG. 1-4). Most walkways are 3 feet wide, ample space for pushing a wheelbarrow or lawn mower

12′ × 20′ = 240 sq ft
240 sq ft ÷ 80 = 3 yards of concrete

Needed - 3 yards concrete
          3 - 12′ 2×4 stringers
          1 - 20′ 2×4 stringer
          2 - 4×4 wet post anchors
             (stirrups)
          3 workers - 1 on chute
          2 on screed

Garage

Note:
remove section
of fence to
let truck
through

Fence

House

←————— 12′ —————→
enough room
for concrete
truck

Sliding door

2″ below mudsill
high point

12′

12′ stringers
20′ stringer

Stirrups
3′ from side
1′ from end

Low point

←————— 20′ —————→

I-I Thoroughly plan your concrete job with rough blueprints. Include measurements, concrete yardage calculations, and any special notes.

I-2 A planter separating the house and walkway. Note how concrete runs downhill from back to front along the block wall.

I-3 The concrete provides a sheltered walkway under an overhead deck while providing maximum yard space for lawn and other landscaping.

I-4 Most walkways are generally 3 feet wide but can be wider to better integrate with existing conditions such as an entrance corridor or covered breezeway.

across. Four feet or more is uncommon but might be necessary for specific needs or landscape conformations (FIG. 1-5). Wide walkways are actually mini-slabs; quite useful for outdoor furniture when used as a front or back porch (FIG. 1-6). In some cases, you might want to break up the plain design of wide concrete walkways by inserting treated, 2-×-4 wood stringers, bricks at specific intervals, or your own personal design.

Narrow walkways of less than 3 feet wide are primarily for decorative purposes and are only functional for one person to walk on at a time. This type of walk is best suited for infrequently used pathways with little traffic.

Add an extra inch for 3-foot-wide walkways. This extra inch will not make much difference in appearance but will come in handy during pouring and finishing. This is because most large concrete tools—a tamp, bull float, and fresno—are 3 feet wide (FIG. 1-7). An extra inch easily accommodates these wide tools so that they can be used down walkway lengths in a single motion as opposed to across walkway widths just 3 feet at a time.

For example, if you form a 3-×-20-foot walkway, it will be much quicker and easier to push tools down the 20-foot run in one maneuver

1-5 Besides serving as functional walking surfaces, extra wide walkways provide plenty of space for benches, outdoor furniture, planters, and other accessories.

1-6 This extra-wide walkway entrance doubles as an attractive and dimensionally symmetrical front porch area. Notice how its brick border and inserts help concrete flat work blend with surrounding brickwork.

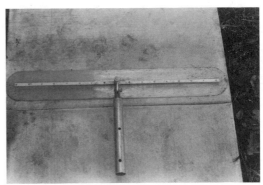

1-7 Consider forming walkways at 3 feet 1 inch so tools such as the fresno can be operated lengthwise.

than to finish concrete across the three foot width, covering only a 3-foot section each time.

Slope

Water runoff is a very important and essential design element for any concrete project, especially when slabs are located near structures. Ideally, concrete walkways and slabs should run downhill toward streets, drains, or natural runoffs (FIG. 1- 8). Proper slope guarantees that water will flow away from certain areas, like next to house walls, and toward specific drainage sites. As you sketch design plans, determine where you will want

1-8 Patio slabs and other projects should slope downhill away from your house and other structures toward drains or other natural water runoff areas.

water runoff to go and incorporate those features into the job. Under no circumstances should you ever want rain or water from any other source to run toward your house or other structure. Always be certain that concrete slopes slightly away from walls and toward the yard or street.

To help you determine the proper slope, use the mudsill on your house as a guide (FIG. 1-9). This is generally a lip located at the bottom of walls where stucco or siding stops near the top of the foundation. This lip should be very close to level. Use it to determine a high point when forming walks designed to run downhill. The space between the top of the concrete and the mudsill will be less at the high point and much greater at the low point, and water will run toward the low point.

1-9 Concrete was poured using the stucco mudsill as a guide. This slab is level under the far wall's lip but slopes down along the left wall.

Many concrete jobs have had to be broken out and replaced because water runoff was not sufficiently factored into the design. Water puddles next to house walls or in low spots around yards negates the primary reason for having concrete walkways, especially during rainy weather. Besides, pools of standing water next to walls and foundations will eventually lead to serious dry rot, mildew, and other dampness problems, especially when water is left in place for prolonged periods.

Along with planning for normal rainwater runoff at ground level, don't overlook runoff from roofs. Investing in rain gutters can solve some runoff problems if they drain away from enclosed planters and into specific drains (FIG. 1-10). Bordered planter boxes without drains can quickly become saturated, accumulating water to pool and flood. Consider installing three-inch diameter drainpipes under walkways to help divert water runoff from roof gutters into drains or toward the street.

1-10 Water flowing out of this rain gutter downspout would surely flood the planter area had a flexible plastic drainpipe not been installed under the walkway.

To prevent mud splatter against your house as a result of rainwater falling from the roof into planters, place concrete close enough to the house so that water drips on the concrete first and then gently falls away. To be sure that concrete is poured close enough to the wall, spray some water on the roof and measure how far from the house the drips land; then form the inside concrete edge accordingly (FIG. 1-11).

PATIO SLABS

Some of the considerations mentioned for designing walkways are also appropriate for patio slabs, especially slope. It is essential that you ensure that the entire slab slopes away from walls and toward open-yard areas and/or drains.

Slope and slab depth are two factors that go hand in hand. Regular patio slabs do not need to be any deeper than 4 inches. Normally, concrete is poured to the top of 2-×-4 forms. When determining a high point for slope consideration, take a good look at the natural slope presented. It is sometimes easier to pour slabs along the natural grade (FIG. 1-12). If, however, the grade is not proper, you will have to remove dirt from the high side to fill in the low side. Dig out only enough to make adequate room for your 2-×-4 form. Excessive digging is unnecessary and only creates more work.

To help determine grade while minimizing excavation work, use 2-×-4 screed boards as guides. Small wood stakes are nailed to the top edge of a

I-11 Walkways poured away from house walls should be positioned so that water drips falling off the eaves land on concrete rather than splattering into planters.

I-12 This patio slab conforms to the yard's natural grade—a downhill slope running toward the bottom of the photo.

2×4 cut to fit inside a formed area. With stakes riding on top of the forms, the body of the 2-×-4 screed will quickly show you when enough dirt has been removed or filled in to make a perfectly even, 4-inch slab depth.

Other features you might consider for patios include open planters, 2-×-4 wood stringer inserts, and means of support for patio roofs (FIG. 1-13). You might even consider pouring two separate slabs of different dimensions connected by a short walkway. For all intents and purposes, any design is good as long as it is functional, channels water away from walls, and meets with your visual approval.

1-13 Small concrete pads were poured outside the slab to accommodate patio cover supports. Installing posts this way provides more open room in the patio area.

MOWING STRIPS

Often, maintaining a well-manicured yard requires some form of separation between grass, flower beds, decorative gravel, mulch, and other landscape features. Lengths of 2-×-4 treated lumber or benderboard work fine for some landscapes, but a more permanent and long-lasting solution is concrete mowing strips.

Mowing strips are generally 4-inch wide concrete curbs poured and finished at ground level or left to stand a few inches above grade depending on their intended function. They can be used to separate gravel driveways from lawns, function as borders around trees, or be placed along fence lines to give lawns a tidy border and make grass edging easier (FIG. 1-14).

Unless there is a special need to partition different ground cover root

1-14 Mowing strips can be rectangularly formed or made to curve around circular driveways and other rounded landscapes. Strips of benderboard are used to form curves.

systems at specific depths, like between Bermuda grass and flower beds, mowing strips do not need to be much deeper than 1 to 2 inches. Mowing strips should be at least 3 to 4 inches thick but you can adjust their tops to be flush with grade level or allow them to stand higher like a curb or raised border. Mowing strips can be formed in rectangular shapes with straight-form lumber or in curved patterns by using thin benderboard forms to match landscaping designs.

Placing concrete inside mowing strip forms is usually done with a shovel. This is because forms are generally not wide enough to allow direct wheelbarrow pours. And, because their dimensions are seldom bigger than 4×4 inches, not much concrete is needed. Small strips may be poured using bags of concrete mixed in a wheelbarrow, while larger projects might be best supplied by 1-yard haulers (described later) or with concrete left over from another job.

EXPANSION JOINTS

Believe it or not, concrete actually expands and contracts with variations of temperature. To help prevent cracks along walkways and slabs, finishers install expansion joints at strategic points (FIG. 1-15).

An expansion joint is created when a piece of special felt, benderboard, or other treated lumber is placed in concrete as a means to absorb the expansion and contraction of concrete as it heats or cools (FIG. 1-16). These inserts are especially important along walkways and at other places where slabs of concrete intersect or butt up to other slabs; like, the edge of a driveway poured up to the edge of a sidewalk.

Expansion joints should be placed about every 10 feet on walkways

1-15 This crack could have been avoided if an expansion joint was installed at the corner just to the right of the post.

1-16 The expansion joint on the right side and the seam running perpendicular to it are correctly installed and should be provided on all walkway corners.

and at every point where corners are formed or intersections occur. When a walkway extends from a patio slab, for example, an expansion joint should be inserted in line with the outer edge of the larger slab. If not, a crack will eventually develop across the walkway at the point where it butts against the larger slab.

Plan ahead for the adequate placement of expansion joints to be sure they are inserted at equal intervals. This ensures that they conform to a uniform pattern. On a long walkway measuring 27 feet, for example, put in a joint at the 9-foot mark and another at the 18-foot mark. Each section will measure 9 feet and look uniform.

Expansion joint felt is available at lumberyards and ready-mix concrete plants (the place where concrete delivery trucks are filled). Although most city building codes require $1/2$-inch felt be placed between sidewalks and driveways and any other city owned concrete slab, you could opt for redwood benderboard or treated 2-×-4 lumber expansion joints on other concrete projects around your house.

THE FINISH

Concrete slabs with a smooth surface, like garage floors, might look nice but are dangerously slick when wet. To prevent slabs from becoming as slippery as ice when wet, finishers enhance traction by applying any one of a number of various concrete finishes. By far the most common are broom finishes, which are very easy to accomplish and hide many finishing errors. Small lines left by hand trowels are quickly wiped away by broom bristles, leaving their own distinctive line pattern. These broom lines not only make slab surfaces look neat and uniform, they add a great deal of positive traction on wet concrete surfaces for pedestrians.

Broom finishes can feature more than just straight line patterns. Brooms that are wriggled as they are pushed and pulled across surfaces create wavy line designs. With a hand-held foxtail broom, you could even design a series of half-circle lines similar to the effects windshield wipers make on auto windshields.

Broom finish texture is determined by how wet concrete is when a broom is applied over its surface. The wetter the concrete, the more defined and deeply textured the lines will be. Conversely, operating a broom over a more cured surface results in very light lines. Heavier-textured concrete surfaces offer maximum traction. Consider heavy lines for steep driveways that experience auto traction problems. For walkways and patio slabs, light broom finishes generally offer plenty of excellent traction for everyday foot traffic, even when concrete is wet.

INTEGRATING DESIGN

Concrete projects can be designed to conform to almost any landscape scheme. If your yard is decorated with lots of shrubs and flowers, you might prefer to leave small areas open inside your patio slab for plants or trellis features. Walkways don't always have to be straight to be functional. Some creative plan sketching might result in a unique blend of offset walkway sections around trees or bushes, adding flair to an otherwise mundane landscape.

Along with the concrete projects featured in this book, a lot of creative ideas are offered through home improvement magazines and other house and garden periodicals. Take a walk through your neighborhood to see what kind of unique concrete designs have been incorporated into landscapes similar to yours. Better yet, you might even have fun driving through exclusive neighborhoods to find truly one-of-a-kind designs that could easily be duplicated or expanded upon.

Because concrete flat work projects are generally considered relatively permanent additions, take the time needed to design a perfect scheme for your landscape. As you work to set forms and grade, don't be afraid to change designs if need be. It is always much better to take a few extra days to plan and form a perfect slab or walkway with exact dimensions than to pour and finish a project with end results that are less than satisfying, either functionally or visually.

Chapter **2**

Concrete costs

Calculating the cost of concrete is not difficult. There are some variables that must be figured in, but once you understand why and how certain fees are charged, you should be able to estimate delivery costs quite closely. After all of the forms have been set, measure the project's total square footage. These figures will enable you to get a firm delivery price, which should only be altered if the concrete truck remained on your job site longer than anticipated.

BASIC COST

Concrete is sold by the cubic yard, broken down into 1/4-yard increments. One yard of concrete generally sells for around $50, depending on the ready-mix concrete company and geographical location. If, for example, you were to order 1 1/4 yards of concrete, the basic cost would be (using a $50 per yard figure) $62.50—$50 for the first full yard plus $12.50 for the 1/4 yard. Chapter 3 explains how to calculate concrete yardage based on square feet.

On top of this basic per-yard cost, there might be extra charges. Short load and overtime fees are common. Actual dollar amounts and time limits vary among companies. Allow for sales tax, too.

Ready-mix concrete companies operate independently and frequently specialize in specific types of work. Some outfits are often set up to supply large-scale builders with hundreds of yards of concrete per week and prefer to shy away from smaller pours. Other companies, generally the smaller ones, are better equipped to deliver smaller concrete loads and are well suited for common do-it-yourself projects. With this in mind, expect larger companies to charge more for smaller deliveries than businesses that commonly deliver to homeowners.

Call a number of ready-mix concrete companies to determine their price structure. A dispatcher will answer your questions and even verify the amount of concrete needed for your job by using the square footage figures you give him. Along with a basic cost, all of the other fees can be openly discussed and calculated. By calling a few companies, you can

3 yards of concrete ordered
Concrete selling for $50 per yard
3 × $50 = $150

Customer orders 3 yards, takes 45 minutes to
unload, and requests a 1-yard cleanup.

Basic charge:	$150.00	for 3 yards
Overtime (15 minutes):	7.50	
Cleanup (1 yard):	$100.00	
Charges with no cleanup and	$257.50	
correct order		
Basic charge	$200.00	for 4 yards
Overtime (15 minutes):	7.50	
	$207.50	

A savings of $50 with correct ordering

2-I Concrete companies allow a certain amount of time to unload concrete. Overtime charges are incurred when unloading time exceeds the time allotted. If extra concrete must be delivered after the initial load, you will be charged cleanup fees.

decide which one can offer the best overall price and service for your project.

SHORT LOAD

Most ready-mix delivery trucks can carry 7 to 8 yards of concrete. To compensate for deliveries that are less than a truck's capacity, companies commonly charge what is called short load fees. Like basic costs, fees vary among companies.

Generally, you can expect to pay about an additional $7 per yard for every yard under seven. If 3 yards of concrete are ordered, it will cost an extra $28 above the basic cost. At $50 per yard, a 3-yard delivery would cost $178—$150 for 3 yards delivered plus $28 in short load fees—$7 per yard for every yard under seven (7 yard-capacity minus 3 yards delivered = 4 yards short load).

Short load fees are not hidden and every concrete company is familiar with them. By shopping around, though, you might find one that only calculates short loads based on 5-yard capacities. In other words, you might only have to pay a $7 short load fee for every yard under five instead of seven.

OVERTIME

Concrete companies allow customers a specific amount of time to unload concrete. After that time has elapsed, they charge overtime. As with short load fees, this is common practice used to compensate for the extra time a truck remains at a job site when it could otherwise be delivering concrete somewhere else. Dispatchers rely on concrete finishers to unload trucks within specific time frames so they can schedule trucks for other deliveries throughout the day.

Some companies allow only 4 minutes per yard unloading time while others might allot up to 10 minutes or more per yard. Some charge a flat per hour fee in addition to actual concrete costs. While gathering concrete cost estimates from ready-mix dispatchers, be sure to ask about their unloading time limits and related fee schedules.

Overtime charges are based on minutes. Rates vary but are generally around 50 cents per minute for every minute over the allotted unloading time. Even though 50 cents a minute might seem like a lot, don't rush your job just to escape this extra fee. An additional 10 minutes at 50 cents a minute will only cost $5, and that might be all the time you need to catch up on screeding or other important chores to guarantee a first-class pour.

A lot of times, concrete truck drivers are in control of overtime fees. In this case, do yourself a favor by getting on the driver's best side by offering him, or her, a cup of coffee or cold drink upon arrival. Fully describe the job layout and point out specific problem areas where concrete placement might be difficult or hard to reach. If the driver recognizes that you are properly prepared, with all helpers and equipment ready to go, you stand a fair chance that a few minutes of overtime might get overlooked.

Getting ready-mix concrete out of a truck and into forms will be the single most physically demanding and important phase of your job. Should you experience problems with concrete placement or screeding, or the project at hand just takes longer than anticipated, don't fret. It is much more important to place concrete correctly the first time than to worry about a few extra dollars in overtime charges.

Concrete is heavy and can quickly tire workers if they have to move it from spot to spot after it's poured; especially to correct mistakes made while hurrying to unload just to beat overtime charges. Your chances of avoiding overtime fees are enhanced when your job site is properly prepared, forms secured and braced as needed, and workers thoroughly familiar with the job before the concrete truck arrives.

CLEANUP

A cleanup load is one that is made after the first truck has made its delivery and there was not enough concrete on board to complete the job. This only applies to jobs where the original truck had the capacity to carry the full amount but the customer made an error and simply failed to order enough concrete. For example, if you ordered 3 yards and that was not enough, the second truck coming with a cleanup load is going to cost a lot. In some cases, just one extra cleanup yard of concrete can easily cost $100 or more (FIG. 2-1).

Contractors frequently rely on cleanup deliveries for extra large jobs when lots of footings, stairs, or other irregularly shaped projects are involved and accurate yardage estimates cannot be fine-tuned to within $1/4$-yard increments. For these steady customers, who order hundreds of yards of concrete a week, cleanup fees are waived. However, should a miscalculation on your part result in a shortage of concrete, you will have to pay a hefty cleanup charge in order to get a second truck to deliver more concrete.

The logic behind cleanup fees is simple. If one concrete truck could have easily delivered 4 yards initially instead of just the 3 yards ordered, it could have been on site a short while, delivered its load, and then been on its way to another job. Since it has to make a special trip back to the plant and then back to your house for a cleanup, however, the company has to rearrange its delivery schedule, pay for extra driver time, fuel, and other overhead. The extra cost and inconvenience is made up through cleanup fees.

Frequently, a few small privately owned concrete companies try to be very accommodating to homeowners and do-it-yourself concrete finishers. They gladly answer questions and some will even visit job sites to offer advice, take measurements, and calculate concrete yardage. This type of personal service could be available in your area just for the asking.

DRIVERS

A reference to concrete truck drivers was made earlier in regard to overtime fees. In addition to that, it is your responsibility to confer with the driver so he comprehends exactly how you intend to pour your job. After all, he is the one who will maneuver the truck and control concrete unloading speed.

Drivers depend upon concrete delivery for a living, and it stands to reason that they have witnessed a lot of different pours. Chances are, they have learned a few tricks of the trade by working alongside or watching professional finishers in action. Make sure they survey your work site. Not only will they get a clear picture of the job for driving and maneuvering needs, but they might also be able to pass along a few helpful hints, such as whether certain forms need extra bracing or how concrete placement could be made easier.

Thanks to power steering and plenty of experience, veteran drivers can often maneuver their big trucks into unlikely tight spots. By getting closer to forms, drivers can eliminate wheelbarrowing, which makes placing concrete much easier. Confer with the dispatcher to determine how much room is needed for trucks. Then, if at all possible, provide the space.

ONE-YARD HAULERS

The cost of ordering concrete from a ready-mix plant for jobs of 2 yards or less becomes quite expensive because of short load fees. In such cases, you might consider using a concrete trailer, commonly referred to as a one-yard hauler. A number of concrete companies, gravel yards, and home-improvement centers rent one-yard concrete trailers for as little as $65 per yard, calculated in 1/4-yard increments.

These heavy-duty trailers are filled with fresh concrete mixed at the rental facility through mini concrete machines. Concrete is very heavy, you'll need a stout pickup truck or utility vehicle to pull the heavy load. Don't rely on small cars with snap-on bumper hitches to work safely. If need be, borrow a friend's pickup truck equipped with a heavy-duty dock bumper.

There are no extra charges for trailer rentals and most companies allow two hours unloading time with no overtime fees. Concrete can be ordered in 1/4-yard increments, perfect for small jobs like mowing strips.

Traveling distance for one-yard haulers is limited to about five miles. As trailers are operated over roads, normal vibrations cause rock and sand to settle toward the bottom. Prolonged travel results in a mix of extra rocky material at the bottom and soupy cream on top. Because additional on-site mixing is not possible, concrete mix consistencies could be jeopardized.

Trailers are unloaded in one of two ways. Some styles are equipped with a small door at the bottom edge of the rear trailer panel. A jack mechanism at the front allows users to elevate the front of the trailer to tilt it so concrete can flow out of the door as it is opened. Others simply have a trough incorporated along the top edge of the rear panel. As the jack tilts up the front of the trailer, concrete pours over the top of the back panel.

If you must travel more than 5 miles with a one-yard hauler, ask the batch plant operator to use the smallest rock (aggregate) possible; pea gravel is best. Smaller aggregate measuring no more than 3/8 inches in diameter (pea gravel) has a tendency to stay suspended longer than larger material that generally measures 1/2 to 3/4 inches in diameter. This should result in a much better overall mix once you arrive at your project site.

Along with prescribed amounts of sand, gravel, and water, sacks of cement (the powder that holds everything together) are added to mixes to make concrete. Unless otherwise requested, ready-mix plants usually add five sacks of cement to every yard of concrete. One-yard batch plants, on the other hand, frequently use just four sacks of cement per yard. Although it might cost $5 a yard more, you should always request at least a five-sack per yard mix. The extra bag of cement gives concrete more strength and makes it more creamy for better surface finishing.

With pea gravel mixes, most finishers prefer six-sack mixes. They have found that this richer concrete blend helps small aggregate disperse and suspend better as well as making finishing surfaces much more creamy. This is a valuable tip for one-yard haulers, especially when concrete is used for flat work projects that will be clearly visible as part of an overall landscaping plan.

ONE-YARD TRUCKS

To complement concrete trailers, some outlets offer 1-yard, mini concrete delivery trucks. A small concrete drum is attached to the frame of a heavy-duty dual-wheel truck. It constantly rotates through a chain-driven hydraulic assembly. Travel distances are almost unlimited. About the only concern is water evaporation from mixes and actual travel time as they relate to mixes setting up and hardening in the drum.

One-yard trucks cost more than trailers. This is simply because maintenance and fuel costs are higher. However, you can still expect to save money on 1- or 2-yard jobs when compared to the cost of a regular ready-mix truck delivery.

If your project requires 2 yards of concrete, be sure to section the job in such a way that the first load will completely fill in a specific formed area and not be disturbed when the second load is poured later. You can install an expansion joint at the junction where the first yard stops and the second one starts or make plans for a planter or other solid division at that breaking point. This is an important consideration, as you will not want fresh concrete to butt against material that has already begun setting up, a perfect condition for the formation of cracks (cold joints).

Chapter **3**

Calculating yardage and arranging delivery

The actual mathematics involved in determining the amount of concrete needed for a project is not difficult. If you can add, multiply, and divide, you've got it made. Concrete is sold by the cubic yard, but calculations to determine the amount of concrete needed for a flat work job is done using square feet equations. Remember the old formula for figuring out square feet? Length × width = area. In this case, it will be, in feet:

length × width = area (in square feet).

The square feet is then divided by a constant to give you the amount of concrete yardage needed.

CALCULATING A CUBIC YARD

One cubic yard can be described as a cube measuring 3 feet × 3 feet × 3 feet. To convert the yardage into feet, multiply the sides by each other in feet. For example,

Side #1 × side #2 (3×3) = 9 square feet.

Take this product and multiply it by side #3 (FIG. 3-1.):

9 square feet × 3 feet = 27 cubic feet.

One cubic yard (a yard), therefore, equals 27 cubic feet. One yard of concrete will cover 27 square feet at a depth of 1 foot (FIG. 3-2).

Because your slab will only be 4-inches thick, as opposed to a full 1 foot, you will have to divide those 27 cubes into smaller ones that measure 4-inches thick. To do this, multiply those 27 cubes × 3; because 4 inches is one-third of a foot (12 inches). Those 27 cubes will now cover three times the area when spread out at a depth of only 4 inches (FIG. 3-3). You can now see that a cubic yard will cover 27 square feet 3 times, which

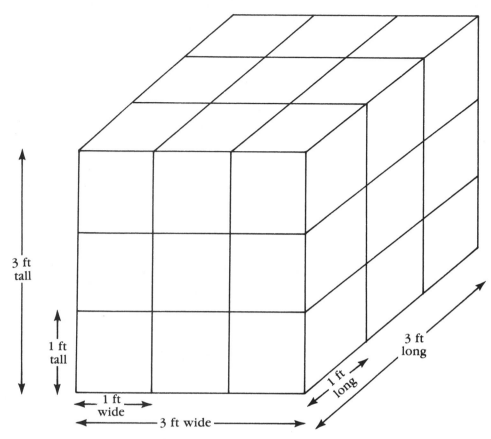

3-1 Equally divided, I cubic yard could be separated into 27 separate cubic-foot blocks; I cubic yard is equal to 27 cubic feet.

equals 81 square feet ($27 \times 3 = 81$). One yard of concrete will cover 81 square feet at a depth of 4 inches.

There are no variations from this basic formula unless the depth of your concrete project is more than 4 inches. So, don't try to squeeze any more concrete out of a cubic yard. If the slab you have planned measures 85 square feet and the depth is exactly 4 inches, one yard will not quite cover it all (FIG. 3-4). In that case, you should order 1¹/4 yards of concrete.

ONE-FOURTH YARD INCREMENTS

Although concrete is calculated by the cubic yard, concrete companies are able to sell it at ¹/4-yard increments. You can order any amount desired, as long as it is totaled in ¹/4-yard segments. For example, if you need extra concrete to cover a slab area of 85 square feet at 4 inches deep, you will have to order 1¹/4 yards. There will be a little concrete left over, perfect for a small mowing strip, trash can pad, or step landing. Just be sure the extra projects are formed and ready for "mud" at delivery time.

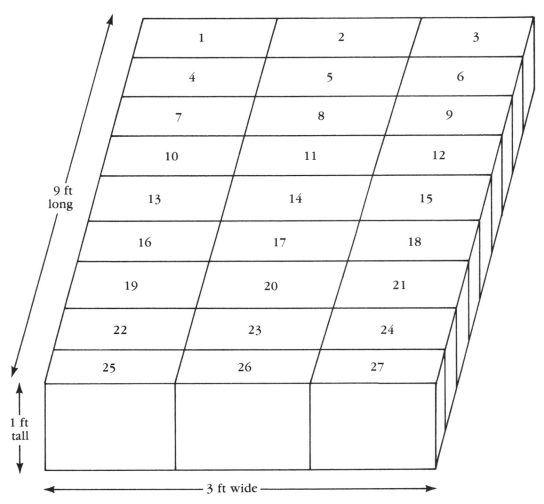

3-2 One cubic-yard of concrete will cover 27 square feet at an even depth of 1 foot.

This same 1/4-yard increment formula holds true for slabs that measure less than 1 yard but more than 3/4 yards. You still have to order the next highest 1/4-yard increment to be certain you will have enough concrete for the job. With that in mind, try to measure out jobs that will most closely engage areas with figures close to 1/4-yard intervals. Tip: 20 square feet at a depth of 4 inches equals 1/4 yard.

Cubic yard shortcut

The number 81, as in 81 square feet per cubic yard at 4 inches in depth, is a difficult number to use as a divider. It will not evenly fit into even numbers, i.e., 200 divided by 81 will not produce an even-numbered answer. You might have a lot better luck using the number 80 instead of 81 (FIG. 3-5).

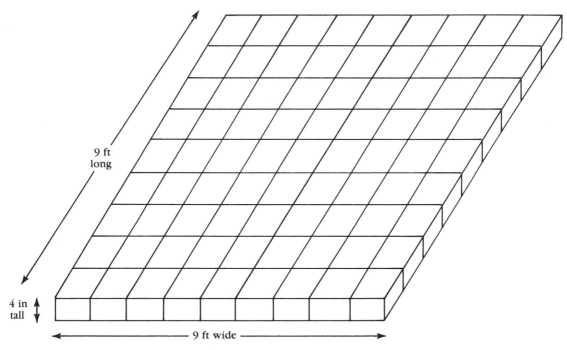

9 ft
long

4 in
tall

9 ft wide

3-3   Concrete flat work is normally 4 inches deep. One cubic-yard of concrete will cover 81 square feet at an even depth of 4 inches.

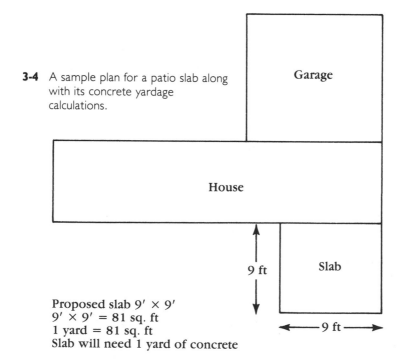

3-4   A sample plan for a patio slab along with its concrete yardage calculations.

Garage

House

9 ft

Slab

9 ft

Proposed slab 9' × 9'
9' × 9' = 81 sq. ft
1 yard = 81 sq. ft
Slab will need 1 yard of concrete

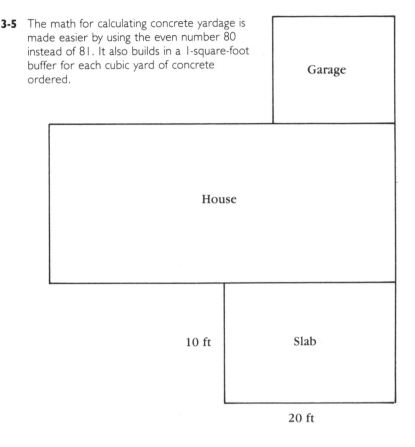

3-5 The math for calculating concrete yardage is made easier by using the even number 80 instead of 81. It also builds in a 1-square-foot buffer for each cubic yard of concrete ordered.

Garage

House

10 ft   Slab

20 ft

Proposed slab 10′ × 20′ - - - 10′ × 20′ = 200 sq. ft.
200 sq. ft ÷ 81 = 2.469 yards of concrete
200 sq. ft ÷ 80 = 2.5 yards of concrete

Using 80 as opposed to 81 has another benefit besides easier mathematics. It gives you a 1 square foot buffer per yard of concrete. By omitting that one square foot, you have gained a little concrete. This very small amount could be a lifesaver on a job that might not be graded to a perfect 4-inch depth or if a helper should accidentally dump a wheelbarrow of concrete on a job you had calculated perfectly to the 1/4 yard.

Whenever calculating yardage for a concrete job, multiply the length times the width and then divide by 80. Use accurate measurements and overestimate rather than underestimate. For example, if a side of a slab measures 9 feet 10 inches, use a figure of 10 feet even. It is much better to have too much concrete than not enough. The same philosophy holds true for actual yardage. If your figures come out to 190 square feet, round it off to 200 square feet. You'll have to graduate up to the next 1/4-yard increment anyway, so why not make your calculations easier while creating a slight buffer of extra concrete at the same time.

Yardage example

Let's do an example for a slab that measures 10×20 feet. To determine the amount of concrete needed, multiply the length times the width, then divide that answer by 80:

10 feet × 20 feet = 200 square feet

This is the total square feet included in the slab. Remember, it must be graded and formed at a normal depth of 4 inches. Now, because this area is 4-inches deep, we divide the square feet by 80:

200 sq. ft. ÷ 80 = 2$\frac{1}{2}$ or 2$\frac{1}{2}$ yards of concrete.

Using this formula ensures that enough concrete is delivered to complete a 10-×-20 slab. Again, the depth cannot exceed 4 inches at any point. This cannot be overemphasized. The vast majority of novice concrete finishers who run short on their concrete loads do so because grading is not accurate. Although their math might have been perfect, irregular depths of 5 inches or more simply used up more concrete than anticipated.

One of the major concerns most novice do-it-yourself concrete finishers have is their lack of confidence in ordering enough concrete to avoid any cleanup charges. If you follow these directions and accurately grade to a 4-inch depth, however, you should not have any problems. Also, be aware that most 2-×-4 milled lumber actually measures 1$\frac{1}{2}$× 3$\frac{1}{2}$ inches, not a full 2×4 inches. Therefore, if you use a normal 2-×-4 board as a guide for grading, you'll really be placing that grade at a depth of between 3$\frac{1}{2}$ inches and 4 inches. Also, slabs of 3$\frac{1}{2}$ inches fair just as well as full 4-inch-deep slabs for normal use, even for driveways.

AGGREGATE

The aggregate (rock) used to mix concrete is available in different sizes and most concrete companies stock aggregate according to diameter size. Although some companies use aggregate as large as an inch in diameter, most concrete deliveries contain material mixed with $\frac{3}{4}$-inch diameter and smaller. Anything larger than $\frac{3}{4}$ inch can pose problems for slabs and walkways that require a smooth, good-looking broom finish. Larger aggregate batches fair better for footings and foundations where material is vibrated with a special tool to provide smooth side finishes. Other jobs, like wall grouting and those that use a concrete line pumper, require pea-sized gravel aggregate.

Professional concrete finishers sometimes special order smooth rock as opposed to crushed rock. Smooth riverbed rock is most common on exposed aggregate finishes where the top layer of cement cream is washed away after the slab has been properly finished and set up strong enough to allow finishers to walk on it without making any marks. After the slab is washed off three or four times and the water is clear, smooth rock surfaces are exposed to reveal a truly custom finish.

If your job is planned as a special project, make sure you discuss your aggregate needs with the concrete dispatcher. He or she will be able to address your needs and recommend the right size aggregate for both visual acceptance and strength requirements.

MIX

Although concrete is sold by the yard, it is actually mixed by weight increments. Specific weight measurements of sand, aggregate, and water are mixed with five to six sacks of powdered cement to make a cubic yard of concrete. Cement is a pure powder, while concrete is the term used to describe the entire finished mixture of sand, aggregate, water, and cement.

Ready-mix batch plant operators use a predetermined chart to gauge materials as they are added together to form mixes of concrete. Sand and aggregate in increments of hundreds of pounds are added to gallons of water and then thoroughly mixed with appropriate amounts of cement.

Concrete batch plant operators can alter mixes as needed. Some jobs might require specific aggregate sizes, others might need heavier sand concentrations. Still, some city regulations might require all concrete be mixed with six sacks of cement per yard. This is common in areas that experience severe winters and other extreme climates (FIG. 3-6).

3-6 On the left is smooth rock aggregate measuring 3/4 inch and smaller. The handful on the right is pea gravel. You can specify either size aggregate when ordering your concrete.

Six-sack cement mixes are also recommended for jobs where concrete line pumps are used to pump concrete from the delivery truck to inaccessible areas. These small trailer-mounted pumps use a small-diameter hose and pea-sized gravel aggregate because larger aggregate will not fit through the small hose. The extra rich and creamy mixture of six-sack per yard concrete makes finishing much easier. A sack of cement generally costs about $5. Expect to see that added charge on bills for deliveries of six-sack mixes.

PLANNING A POUR

The overall type of concrete pour you plan will determine the concrete mix and the amount of help you'll need. Helpers are an integral part of any job. Professionals can handle small pours all by themselves because of their experience and expertise. Novices should plan on having some help available, however, no matter how small the slab or walkway.

If you need to use wheelbarrows to get concrete from a truck to the forms, have a number of helpers on site. For a 2- to 3-yard job, it is recommended you have at least four, preferably five, workers. You'll need two people to run wheelbarrows, two to screed and one to help move concrete on the ground to assist the screed effort. If your job requires more than three yards, plan on having one or two more helpers; at least for the amount of time it takes to unload concrete.

Regular, five-sack, 3/4-aggregate concrete is fine for almost any walkway, patio, or driveway. It can be wheeled with no problem and it finishes just fine. Other mixes needed for unique design characteristics, or because of local regulations, must be special ordered.

Should your forms be located in such a way that a concrete delivery truck can dump directly into them, you'll still need a few helpers. One to work the truck chute, two to screed, and another to assist the screed work by moving concrete around as needed with a heavy-duty rake or square-point shovel.

When pouring directly from delivery trucks, you must account for the driver's ability to maneuver. Drivers commonly sit in the cab and move the vehicle forward as directed by the chute man while operating the controls that start and stop the flow of concrete down the chute. Most trucks are 7- to 8-yard rigs. Although they are big, you will be amazed at how some veteran drivers can maneuver them into very tight places.

If your job site appears big enough to accommodate such a vehicle but you are not quite sure, call the ready-mix dispatcher and ask about the dimensions needed for minimal truck maneuverability. It may be well worth the effort to remove a section of fence or temporarily relocate a storage shed in order to get a concrete truck close enough to a job site and eliminate the need to wheelbarrow concrete, especially if your job requires more than 2 to 3 yards.

Larger concrete trucks are also available. These rigs can haul up to 10 yards of concrete at one time. The added weight of 3 yards of concrete requires them to have an extra set of wheels. These wheels are hydraulically raised and lowered. They are placed on the ground while the truck is on the road with more than 7 yards. When it arrives at a job site, the wheels are raised out of the way. Even with the wheels raised, this particular truck is very long. Make sure there will be enough room for rigs of this type to operate around your job site.

Determining when to pour

The decision as to which day to pour your concrete job depends on a few factors. Initially, you must determine when all forming chores will be complete. You must then consider the availability of helpers and the concrete supply company's delivery schedule. Coordinating and confirming these factors is critical. If the forms are not ready, you'll either have to postpone the job or pay standby charges while the concrete truck idly waits for you to fix them. In addition, failure to have enough helpers available could make your pour a painstakingly difficult operation.

Concrete delivery companies regularly schedule jobs as they are called in. Certain contractors frequently have standard delivery times scheduled weeks in advance. Around these mainstay contractor customers, concrete companies fill in schedules with deliveries to other customers. Call the concrete dispatcher with your order a minimum of two days before you want delivery. Optimally, you should schedule delivery one or two weeks ahead of time to be certain the date and time you desire is open and available. By confirming a date and time for concrete delivery this far in advance, you have the best chance of securing the date and time most suitable for you, as opposed to being scheduled in at the concrete company's most convenient time.

For the most part, weekday deliveries are best. This is when concrete plants are operating with full crews. Many companies make Saturday deliveries but commonly charge additional fees to cover driver and dispatcher overtime wages. Fees can range from $5 per yard to flat rates of around $25. Weekend delivery charges are not standard and vary widely from company to company.

Determining the time to pour

Concrete hardens as water and moisture leave the mix. This occurs through moisture absorption into the ground and through surface evaporation. During hot weather, water can quickly seep into parched dirt (the main reason formed areas are always sufficiently wet down before pours) and evaporate at a tremendously rapid pace. When this happens, slab surfaces dry too fast, which can result in cracks or missed opportunities for workers to effect attractive surface finishes.

Conversely, cool winter temperatures combined with seasonally soggy soils cause concrete mixes to retain water and moisture for extended periods of time. Under these conditions, water has no place to go. The ground might already be saturated from rain and cooler temperatures will hinder normal surface evaporation.

Most professional concrete finishers prefer to pour concrete as early in the morning as possible. On long, hot summer days, pouring in the morning allows concrete to harden at a slower rate during cool morning temperatures as opposed to quickly setting up during the heat of the day. During winter, concrete takes much longer to set up and you'll need as much daylight as possible to effect final finishing operations. Pours between 7:00 AM and noon generally work best. Some concrete companies, at a customer's request, will use hot water in the mix to speed curing times.

Generally, figure that concrete will take about four to five hours to

cure enough to support final finishing maneuvers during mild temperatures between 65 and 75 degrees Fahrenheit. Cooler temperatures may
require more, while hot days exceeding 90 degrees Fahrenheit could
cause concrete to flash in less than an hour. During winter, many concrete
workers have had to set up temporary lighting to see well enough to finish
slow drying concrete late at night. While during hot summers, they have
had to work at feverish paces just to be able to get a halfway decent finish
on fast curing concrete surfaces. In both situations, early morning concrete pours provide the best opportunities for optimum concrete curing
conditions and surface finishing efforts.

Inclement weather

Concrete cannot be adequately finished in rain or snow. Raindrops and
snowflakes disturb finishes by adding water to already wet surfaces and
diluting the creamy top texture. Unless you have a means of covering
slabs with a tentlike structure for protection against the threat of rain or
snow, cancel the delivery and reschedule. Concrete dispatchers do not
like cancellations but are accustomed to it during bad weather periods.

Professional concrete finishers have been forced to cancel and
reschedule concrete pours as many as five and six times due to inclement
weather conditions. Although this causes scheduling nightmares, it is far
better to be safe than to risk losing a slab finish to the pockmarks, holes,
and uneven finish left behind by raindrops and water runoff. Just remember, once concrete is on the ground, you have to stay with it until the surface is finished and hard enough to withstand raindrops without being
adversely affected.

Never be afraid to cancel a concrete delivery. Dispatchers are used to
these problems and will reschedule at your request; just don't wait until
the last minute to cancel. Sometimes, especially when a lot of cancellations have occurred, trucks will run ahead of schedule and could show
up at sites an hour early. Once a truck arrives at your job site, you will be
expected to unload it, or, at the least, pay a delivery charge even though
no concrete was poured. Therefore, determine as early as possible what
the weather will be, and call the concrete dispatcher immediately if you
decide to cancel.

ORDERING

The first thing a dispatcher will want to know about your job is what type
of mix is needed—$^3/_4$, five-sack, pea gravel, six-sack, smooth rock, etc.
You'll then be asked how many yards are required. If you are not quite
sure that your yardage calculations are entirely accurate, give the dispatcher the dimensions of your job and compare both yardage results. If
both figures match, you are in business.

Next comes actual scheduling. Once you request a particular date
and time, the dispatcher will check a master schedule to determine how
your order fits in with other orders on that day. Enough time must be
allotted for driving time both to and from your job site, as well as actual

unloading time for the amount of concrete ordered. You may be asked to push your delivery time ahead or behind in order to best fit within the master schedule. This is why calling a week or two early works best—because you may be first on the list instead of just a late order used to fill in an otherwise full calendar.

Inform the dispatcher of any special conditions pertinent to your job. Is the site address easy to find or is it off a dirt road out in the country? Will there be any unusual obstacles for the driver to maneuver around? Do you need concrete delivered on the dry side because you are pouring a steep driveway? Factors like these are important and the more you explain your pour to the dispatcher the better your chances are of receiving a timely delivery with excellent service.

Predelivery call

On the morning of your scheduled concrete pour, plan to call the concrete dispatcher a couple of hours before delivery to confirm your order. This helps both you and the dispatcher. By confirming your order, you let the dispatcher know your job site is prepared and the pour should operate in a timely manner. This information assists in overall scheduling and adds an element of on-time service for customers. For you, this call confirms that the address, mix, yardage, and scheduled delivery time are all as expected. You might find that previously delivered orders are ahead of schedule and yours can be brought out earlier if desired or that complications have developed and deliveries are running late.

Chapter **4**

Grading and forming tools and materials

Pouring a concrete slab or walkway consists of two basic phases, grading and forming then pouring and finishing. This chapter focuses on the tools and materials required to achieve the first phases.

GRADING

The term *grading* refers to work directed to leveling the ground (base) for a slab to be poured on top of it. A flat base offers the most uniform support for slabs and deters cracks. Flat and even base preparations also ensures accurate dimensions for concrete yardage calculations. Most of the tools needed for grading and forming operations are common household items that most do-it-yourselfers already have on hand.

Heavy-duty shovels are a must. Both round and square points have specific uses, although the majority of work is best accomplished with a square-point shovel. Its straight edge is handy for squaring graded side sections and scraping bases down to grade. A round point shovel is best for digging out trenches and other semi-deep appendages for drains, electrical lines, water pipes, and other items.

A heavy-duty garden rake can be quite useful for smoothing out small mounds of dirt inside formed areas to effect a final grade. It can also be used to gather rocks and other debris strewn about otherwise-even bases. This type of rake also works well in moving wet concrete during screeding.

Unless the ground you are going to grade is easily broken loose and moved around with just a shovel, you will need a sturdy pick. A pick with a point on one end and a wide blade on the other works best. The point works great for breaking up hardpan material while the blade makes quick work of chipping away high spots and cutting out side sections. Hoes are also handy for removing loose dirt or chopping out grassy sections.

The most important and greatly used tool in concrete construction is a heavy-duty contractor's wheelbarrow with an air-filled tire. This is the best and most versatile wheelbarrow for concrete construction. Lightweight garden wheelbarrows with solid rubber wheels cannot stand up to the heavy burden of concrete hauling. If you do not have a heavy-duty wheelbarrow, plan to purchase one at a lumberyard, hardware store, or rent one from a rental company. If your grading job does not require a lot of dirt removal, you can certainly get by without one, but you should have one available for the actual concrete pour; up to three if your job calls for concrete to be wheelbarrowed from the delivery truck to the forms.

Grading can be either a backbreaking chore or a rather simple operation depending on your concrete design, existing landscape grade, and base material characteristics. For the most part, it is easiest to form jobs before grading them. Not only does this clearly distinguish the job perimeters, it also sets grade requirements at the same time, giving you an accurate means to gauge grade depth and dimension progress. Chapters 6 and 7 contain more in-depth information on grading techniques.

FORMING TOOLS

Setting concrete forms consists mainly of carpentry work—sawing, nailing, bracing, and supporting board lumber—and requires an assortment of carpentry tools such as a claw hammer, sledgehammer, tape measure, level, handsaw (power circular saw optional), square, pencil, chalk line, string, nail bag, and other tools.

Claw hammers and sledgehammers

A claw hammer is used to drive nails through support stakes and into forms, as well as for other nailing chores. Although you might be able to get by driving stakes into the ground with a claw hammer, a three-pound mini-sledgehammer works much faster and more accurately. It also helps to prevent wood stakes from splitting after repeated hammer blows. The wide-striking face and added weight of hand-held sledgehammers are more suitable for securing stakes.

Tape measure

A heavy-duty, inch-wide metal tape measure with a length of 25 feet is just about perfect for most concrete forming jobs. This type of wide metal blade extends at least 7 feet in midair without bending or bowing, an ideal attribute when measuring distances where there is no means to hook the blade's end clip. For extra long walkways measuring more than 25 feet, consider using a 50- or 100-foot tape measure.

Saws

An ordinary handsaw is generally adequate for most forming needs. At the most, you should only have to cut the ends off a few forms to make everything fit properly. If desired, you can even cut the tops off wood stakes to make your screeding job progress a little smoother. Be sure the saw you use is sharp. Dull saw blades make difficult work of cutting straight lines.

Square and string

Unless your concrete design consists completely of curves and rounded features, you'll need a square to make sure corners are set at 90 degrees and forms come away straight from structures such as your house or garage. A small square is also useful as a guide for making straight lines on boards for cutting guides. However, most professionals prefer large squares to accurately gauge corner dimensions and other perpendicular attributes.

Heavy-duty string is a valuable asset in concrete forming, especially while forming long walkways. String is stretched taut between two secured stakes to serve as a perfectly straight guide for installing form lumber. String can also be used to help set up a forming job before forms are actually placed on the ground. By systematically placing stakes and stretching string between them, you can get an excellent idea of how a concrete design will look and function in relation to the existing landscape.

Levels

A string level is an inexpensive and handy tool that can be attached to a section of heavy-duty string or line. (See FIG. 4 1.) Used in conjunction with a tightly stretched string that serves as a form guide, an attached string level shows grade slope. The string can be raised or lowered at either end until it perfectly marks the point where top form edges should be positioned. A taut string signifies a straight line for form installation as well as marking its height position.

Standard carpenter levels are used to set grades for medium length forms measuring up to about 20 feet while a string level is used for walkways in excess of 20 feet. A 2- to 4-foot-long carpenter's level would be better suited for forming patio slabs, driveways, and other small- to medium-sized jobs. Professional concrete form installers rely on 4-foot levels for more accurate readings. This is especially important when forming house slabs, stem walls, and other construction projects. Professionals then use transits for precise measurements and level readings.

When forming small to medium slabs, use, as a minimum, a 2-foot-long level. Although small torpedo levels might be fine for really small work around the house, they are much too short to offer viable readings when working with long form lumber.

4-1 A carpenter's string level helps form installers establish grade. Be sure string is stretched tight to ensure accurate carpenter's string level readings.

Chalk line

A chalk line, also called a snap line, is nothing more than a roll of heavy-duty string secured inside a small metal case filled with regular, powdered chalk. It is used to mark straight lines along house foundations, walls, and other stationery items. A chalk line and powdered chalk are available at lumberyards, hardware stores, and other tool outlets.

With the end of the string secured on a nail or held tight by a helper, you can stretch from one designated marking point to another and then gently lay the string down until it touches the surface to be marked. Then, simply pick up the string, pull it an inch or two away from the surface, and let it snap back. The result will be a perfectly straight chalk line that can be used as a guide for forms.

Other tools

Two miscellaneous items you might also consider purchasing are a heavy-duty carpenter pencil and a nail bag. Heavy-duty carpenter pencils are rugged and hold up well in construction work. The thick, wide pencil lead is strong enough to hold up without breaking while marking lumber, walls, foundations, and other objects during forming. Unless your concrete job is very small and you have no intention of ever doing another outdoor construction project, you might seriously consider purchasing a nail bag set for nails and tools.

FORMING MATERIALS

Basic concrete forming material consists of 2-×-4 form lumber, wood or steel stakes, and double-headed (duplex) nails. This lumber is strong and its size is just right for normal concrete slab depths. Although steel stakes will virtually last forever, their cost is not justifiable for infrequent do-it-yourself concrete finishers. Wood stakes are economical and common at all lumberyards. Be sure to buy several more than you need, however, because many will split or break during forming operations.

Double-headed nails feature a large head about 1/4 inch ahead of the normal nail head at the end of the shaft. These nails are driven into stakes until the first head makes contact with the surface. When the concrete job is complete, the end head sticks out far enough so that it can easily and quickly be grabbed by the claw on a hammer and pulled out.

When selecting form lumber, look carefully at a board's straightness, the most important facet of form lumber. Crooked forms are difficult to straighten, even when pushed and pulled and secured with stakes. And, if your forms are installed crooked, your walkway or slab edge will be crooked, a true sign of an amateur concrete job. Choose each form board separately and look down its length for any signs of bows, twists, or bends. Be very selective and purchase only those that are as perfectly straight as possible.

Lumber is sold in even lengths: 6, 8, 12, 18, and 20-foot boards. Anything longer has to be special ordered and might not be as easy to work with or as straight as desired. If the dimensions of your slab total an odd number of feet, like 9 feet for example, you will have to buy a 10-foot board. You might be able to form it in such a way that the extra foot does not have to be cut off and the full 10-foot board can be used for another project later on.

The longer a form is, the more apt it is to be warped. A 20-foot board is much more likely to have bows in it than a 12-foot piece. Therefore, if your job calls for a 24-foot form board, use two 12-foot boards instead. They can be secured together with a 2-×-4 brace on their backside to extend a full 24 feet. Because they are relatively short, chances are very good that they'll be straight and easy to work with.

Knots

Knots on the sides and edges of form lumber can cause imperfections in concrete finishes (FIG. 4-2). Concrete flows into recessed knot holes and then hardens. Don't purchase boards riddled with knots and deep gouges. If form lumber is perfectly straight with just a few knot holes or gouges on just one side or along only one edge, however, you can use it as long as the bad side or edge is positioned away from the concrete.

Large knots that encompass an entire width section create weak spots in a board. Too much pressure exerted against these wide knots will cause boards to crack and separate. If at all possible, hand select your form lumber and avoid boards with large knots, deep gouges, and other obvious imperfections.

Stakes

Inexpensive wood stakes are sufficient for most jobs and can be found at lumberyards or garden centers. They are sold in bundles of 12- to 24-inches long. It is recommended that short stakes be used in very hard ground, like hardpan, and longer, 24-inch stakes for softer soils. In very hard dirt, you might have a difficult time driving stakes more than a few inches deep. In this case, you can augment form support by pushing dirt up against them. In soft soil, longer stakes will easily be driven deep into

4-2 Knots on form lumber, like this one, are weak spots. Be sure to place a stake next to them for added support. Open knot areas on the 1¹/₂-inch side of form lumber must be placed on the ground with only the good 1¹/₂-inch side facing up to give concrete edges full form support.

the ground. This is necessary because soft soil will not support stakes that are only partially driven in.

Have plenty of stakes on hand for any concrete forming job. Although your forms will only be holding back about 4 inches of concrete, you would be amazed at the amount of pressure they must hold up against. Once a form breaks and concrete begins to flow out of the break, you will have to stop all pouring operations until repairs are completed. This costs valuable time and quite possibly extra standby fees.

Be prepared to place a stake every 3 feet or so along forms to guarantee that none bow, are pushed out of position, or break. Never place stakes any more than 4 feet apart, regardless of the circumstances. If a wood stake splits on the hammer blow, plan to back it up with another stake or simply place another one alongside it.

In areas where the ground is exceptionally hard or rocky, you might not be able to use wood stakes. Because of the soil and rock conditions, wood stakes might break every time you try to drive them in. In that case, you might have to purchase steel stakes, rent them from a rental yard, or make do with sections of rebar cut to stake size.

Steel concrete stakes can cost from $2.50 to as much as $5 each or more, so in lieu of purchasing expensive steel stakes for just a single concrete job, consider renting as many as you need for as long as you'll need them, perhaps two or three days.

If you cannot locate stakes to rent and new ones are not within your budget, you might be able to adequately support forms by using rebar cut to sections of 1- to 2-foot increments. Although you will not be able to put a nail through rebar stakes, you can nail next to them and bend nails over their shaft for support. Rebar is sold at lumberyards and other construction supply outlets. It is the type of rod you see sticking out of chimneys and block walls as they are being constructed. Rebar is also widely used as an inner grid-work support for concrete retaining walls and other concrete structures.

EXPANSION JOINTS

Expansion joints are complete separations in concrete walkways and other flat work where a type of compressible material fills the gap to absorb the pressure of concrete expansion and contraction. Concrete expands during hot weather and contracts during cold weather. If flat work is confined on all four sides without some type of expansion release, it will crack. Patio slabs and other flat work open on two or three sides are not confined and can generally expand without cracking.

Expansion joints are usually placed about 10 feet apart and are most common in walkways and sidewalks. Although walkways can easily expand sideways, they are definitely confined lengthwise. Expansion joints are also placed at every walkway corner and where walkways butt up against other slabs, such as driveways. If expansion joints are not installed between separate slabs every 10 feet or so along walkways and at every point where walkway corners or bends are featured, cracks will soon appear.

Expansion joint felt

Expansion joint felt is a specifically designed material made of a semirigid felt board treated with a petroleum sealer (FIG. 4-3). Just about every city or town requires felt be used whenever a private slab or walkway butts next to a public sidewalk or other concrete slab. Although the material works excellent for expansion joints, it is rather difficult to work with.

4-3 Expansion joint felt comes in various widths. This one is 3¹/₂ inches wide. A bundle of rebar reinforcing rod is resting on the ground under the felt strip.

Felt pieces are placed inside forms before an actual pour, then braced with either a 1×4 or 2×4 to support them perpendicularly. After concrete has been poured around the felt, the braces are pulled out.

Using an edging tool next to felt presents a problem. Normally, edgers have a somewhat sharp leading edge that frequently digs into soft felt material that not only makes a mess of the concrete edge but ruins the crisp felt edge. One way to avoid this problem is to place felt strips a little higher than the top of your concrete. You will have to work around it but edging will be much easier. After the concrete has fully cured, about two weeks, use a sharp razor knife to cut felt down to an appropriate level.

Felt alternatives

If expansion joint felt is not specifically required for your concrete slab or walkway, you might consider using an alternative material that is easier to work with and might be more eye-appealing.

A clear piece of 2-×-4 redwood or pressure-treated wood will serve as an adequate expansion joint. It is easier to install and you will not have to lift it out of the wet concrete as you would a brace for felt, it is easier to edge against, and it might look better once it is painted or stained to match the color of your house. Only redwood or pressure-treated material can be used. Untreated wood will rot in a very short time and be almost impossible to replace.

Another easy-to-install alternative is redwood benderboard. This material is common around flower beds as a border to separate grass from other landscaping materials. It comes in lengths of up to 10 feet, widths of up to 4 inches, and is about 3/8-inch thick. As opposed to bracing an entire 4-inch-wide piece inside a walkway form, you can simply cut to whatever length is needed and then split benderboard into 2-inch strips. After concrete has been poured, tamped, and floated, wriggle the 2-inch section into place as marked by nails or other obvious designations. You might have to lightly tap it into place. Afterward, a hand float or trowel is used to smooth concrete around the strip, and an edger maneuvered to form a pleasant rounded corner next to it.

This 2-inch strip of benderboard will not completely separate concrete but it will absorb expansion movement to the point where concrete will crack under it but not on top. This option is quick and easy to use on walkways up to 3 feet wide where underlying cracks will not be noticed or cause problems; like when walkway sides are covered by dirt or lawns. Walkways greater than 3 feet in width are too wide for the acceptable use of benderboard strips, as the wood is too flimsy at that length to be able to be inserted in a straight line. Use instead either a section of felt or treated 2-×-4 material.

OTHER FORMING MATERIALS

Double-headed duplex nails are mainstays for concrete form installers. The second head makes pulling these nails from forms a snap. Nails are available in various sizes—6d, 8d, 10d, and sometimes larger. Eight penny, or 8d, is well suited for most common 2-×-4 forming work. They are long enough to go through wood stakes and almost completely through the sides of 2-×-4 forms to offer plenty of strength and stability. 6d nails are a little too small to be used with wood stakes and 2-×-4 lumber, although they might be fine for relatively thinner steel stakes and 2×4s. 10d nails are just too big for most regular forming chores.

For ordinary forming work, nails are first driven through stakes and then into form boards. This is because stakes are placed along the outside face of form lumber. If these stakes were placed and then nailed from inside the forms, you would not be able to remove them once concrete hardened. One exception is when you must nail from the inside using 2-×-4 forms, such as when you are trying to secure a form located on top of a piece of existing concrete.

For example, let's say you want to pour a slab off the back of your house. A sidewalk runs along the side and will be next to the new slab, but you want the new slab to be 4 inches higher than the walkway. How will you place a 2-×-4 form on top of the existing walkway and secure it? Place stakes right next to the walkway—the exact place the side of the new slab will be. Place a 2×4 next to the stakes and on the outside (the stakes will actually be inside the forms). With the form in place, drive a 16d sinker nail through the stakes and through the form and then bend them over on the back side of the form (FIG. 4-4).

The bent nails will hold the form in place. After concrete has cured enough to support final troweling, straighten the nails. Gently pull the form away from the stakes. The stakes will be embedded in concrete on three sides. The only side bare will be the one against the form. Pull the stakes out of the concrete and repair the areas exposed by the pulled stakes.

Drainpipes

Plastic drainpipe is a very good material for directing water runoff. During the forming process, provisions should be made to insert this pipe where needed. Flower beds located between walkways and house foundations could flood during heavy rainstorms. Rigid or flexible plastic drainpipe installed under concrete as an escape for water accumulation would prevent this.

Drainpipe can also be used to redirect rainwater runoff that is blocked from its natural path, such as a new storage shed slab between your house and the fence line. How will water, which normally drains from back to front, get through the slab to the street? If this is not accounted for, water could puddle next to the slab, possibly flooding the storage area. By simply placing a trap next to the slab with a piece of drainpipe attached, water would easily be able to flow away.

Plastic drainpipe is inexpensive, long lasting, and readily available at lumberyards, hardware stores, and garden centers. Be sure it is completely buried beneath concrete so that it does not extend into the slab or walkway. If positioned high into the flat work, a weak spot will develop

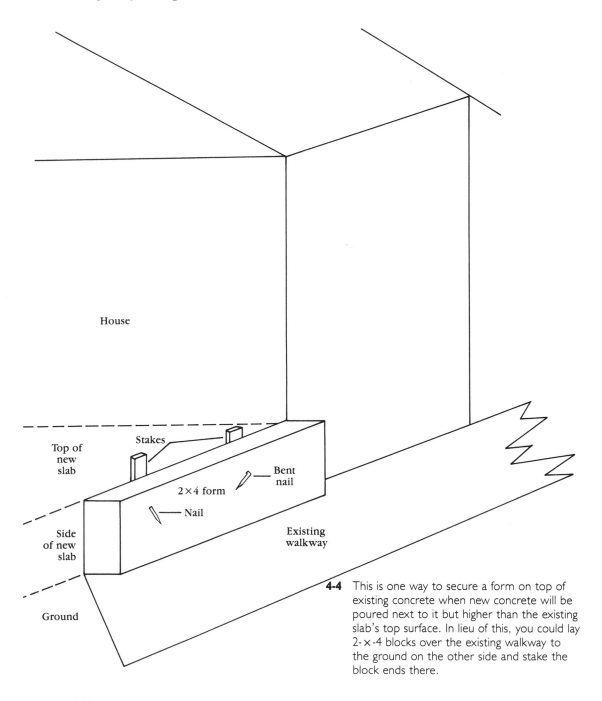

House

Top of
new
slab

Stakes

Side
of new
slab

2×4 form

Nail

Bent
nail

Existing
walkway

Ground

4-4 This is one way to secure a form on top of
existing concrete when new concrete will be
poured next to it but higher than the existing
slab's top surface. In lieu of this, you could lay
2-×-4 blocks over the existing walkway to
the ground on the other side and stake the
block ends there.

that could lead to cracks because the actual concrete will not be its prescribed 3½- to 4-inches deep at that spot. Dig an adequate trench for the
pipe. Be sure that trap connections are tight and secure and the outlet
opening is clear.

Chapter *5*

Pouring and finishing tools and materials

The equipment needed for pouring and finishing concrete ranges from ordinary sturdy shovels and rakes to specialized items like tamps (jitterbugs) or fresnos (3-foot-wide steel trowel with a handle). Although specific, heavy-duty concrete tools will generally last a lifetime, they need not be purchased. Most rental yards keep plenty in stock and loan them out for very reasonable fees. You can purchase your own set of concrete pouring and finishing tools from a company that specializes in such equipment or at some lumberyards and hardware stores. Prices vary, so shop around (FIG. 5-1).

RAKE AND SHOVEL

Because concrete is such a heavy material, flimsy garden rakes and shovels will not hold up. Plan to use only heavy-duty tools built to withstand construction-type work. You will need a sturdy rake to push and pull concrete toward and away from screed boards during pours. Shovels can be used for this chore, but rakes move more material faster.

Square-point shovels are used for a number of concrete pouring jobs. Although round-point shovels do a good job of scraping concrete out of truck delivery chutes or holding it in place along chutes, they allow too much excess concrete to spill out when trying to place concrete inside narrow forms. Square-point shovels work well to not only place concrete inside narrow forms, but to scoop concrete from the ground and fill in low spots.

WHEELBARROW

Using a wheelbarrow to deliver concrete from a delivery truck to forms is hard work. To help this tough job roll along as thoroughly as possible, use

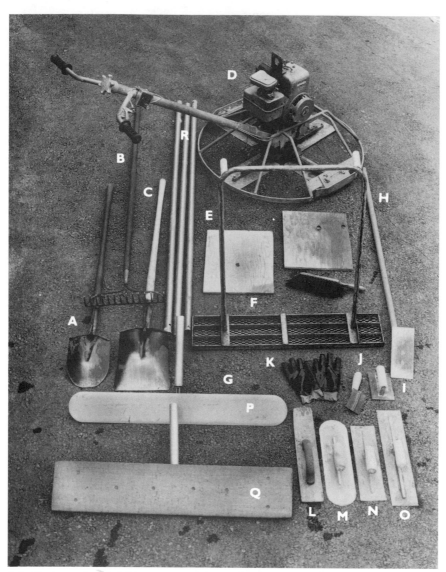

5-1 Tools needed to finish concrete: (A) round-point shovel, (B) rake, (C) square-point shovel, (D) [optional] power finishing machine, (E) knee boards, (F) fox tail broom, (G) tamp, (H) walking edger, (I) hand edger, (J) seamer, (K) rubber gloves, (L) hand float, (M) pool trowel, (N) scrubber and leaner trowel, (O) finishing trowel, (P) rounded fresno, (Q) bull float, (R) extension poles.

only heavy-duty contractor-type wheelbarrows. Smaller, garden-variety types with solid rubber wheels are not designed for concrete loads. Filled with "mud," they are very difficult to maneuver and tend to spill material while being "wheeled."

Construction wheelbarrows are reinforced with metal uprights, have thick, long wood handles for strength and maneuverability, and are designed to move about without spilling loads. Construction wheelbarrows are also built with a braced front nose piece that functions as a stop against the ground when they are tipped upright to dump loads over forms and into slab areas (FIG. 5-2). They are readily available at most rental yards. New ones, purchased from a lumberyard or concrete supply outlet can cost up to $100.

5-2 A heavy-duty contractor's wheelbarrow with an air-filled tire is the best for concrete work. Long handles and a sturdy front nose piece help "wheelers" maneuver and dump concrete loads confidently.

If you have never wheeled concrete before, you must start out with your wheelbarrow only filled part way; no more than half its capacity, even with construction-type wheelbarrows. Although it may not seem like much material, you will be surprised at how heavy the load is and how much physical effort is needed to push the wheelbarrow, turn it around corners, and then tip it up to empty. As you progress through your job, expect to gradually have the concrete truck driver fill your wheelbarrow more and more as you get accustomed to the work. Never fill it to the brim; it is just too heavy to manage.

SCREED BOARDS

Screed boards are straight, 2-×-4s used to level fresh concrete just after it has been placed inside forms. The bottom edge of a screed board is usually laid on top of forms and then pulled across them to sweep back piles of concrete sticking above forms. They are also used to find low spots where more concrete is needed to fill up a form. In essence, a screed board's bottom side determines a slab's top surface level.

Concrete finishers operate screed boards in a sawing-like motion across the tops of forms. This action works well to help aggregate (rock) settle into the surface mixture and move material back while flattening surfaces to a level consistency.

Screeding maneuvers are hard work, perhaps the physically most demanding part of any concrete job. It is also one of the most integral parts of the entire concrete finishing process. Poorly screeded slabs and walkways frequently have uneven surfaces riddled with low spots that hold puddles of water and look terrible.

By all means, take whatever time necessary to ensure perfect screeding during all parts of your pour. If necessary, stop everything else and make the concrete truck sit idle until your screed workers catch up with the concrete that has already been dumped. Should you decide to ignore the importance of screeding and try to catch up later, you'll find that an enormous amount of work will be needed to move excess concrete from high spots and then over to the low ones. While you are busy doing that, concrete will be setting up and you will lag further and further behind on other work that needs to be done.

Most professional concrete finishers realize the hard work needed to accomplish first-rate screed jobs. They also feel that once screeding has been completed, the majority of work is over; at least the hard work that demands most of their physical exertion.

TAMP

Tamps are metal tools commonly referred to as *jitterbugs*. They consist of a metal screen mesh that measures 3-feet-long on the bottom and about 8 inches in width. Attached to each end of the bottom screen are metal arms that reach up to a connecting bar featuring two handles. This tool is used to tamp (push down) rock and aggregate on concrete surfaces in order to bring up creamy textures suitable for finishing. Slabs that are not tamped will have extra rough and rocky surfaces and lack the amount of necessary cream for smooth finishing endeavors.

Most concrete jobs only require tamps be raised a few inches off the surface and then allowed to fall down flat onto the mud. A lot of forceful downward pressure is not necessary. The screen size is just right to force rock down and allow creamy mud to filter through. Results should be relatively smooth surfaces with rocks pushed down at least $1/2$ to 1 inch below a layer of cream. Tamps leave their own pattern over an entire slab surface but they can be quickly and easily wiped out with a bull float.

Tamping can be a very messy chore. Each time the tool comes in contact with fresh concrete, cream will splatter. Expect your boots and pants to get covered. When pouring next to, or near, your house, garage, fence, or other structure, be sure to erect some type of a shield against concrete splatter. Before pouring begins, use strong duct tape or heavy-duty masking tape to secure plastic or newspaper against walls and other things you want protected. Generally, you only need to place protection up about 3 feet from the slab's surface. Should tamping splatter occur higher than that, you are tamping much too hard.

Be sure that newspaper or plastic is not positioned so low as to interfere with pouring or screeding operations or that it actually comes in contact with a slab's surface. The protective material you put up can be removed while you are out on the slab doing the finish trowel work. At that time, should any concrete residue fall off the newspaper or plastic, it can be easily trowelled into the surface.

BULL FLOAT

Screen mesh patterns left behind by tamping are smoothed with a bull float. Although metal bull floats are available, many professional concrete finishers use bull floats made of wood. Essentially, bull floats are 3-foot-long boards, about 8 inches wide and 1 inch thick, equipped with a metal support and a swivel connector to which extension poles are attached because you cannot walk on slabs at this point (FIG. 5-3).

Extensions (poles) are attached to bull floats to reach hard-to-reach places without having to step in concrete. Extensions are usually 6-foot-long poles with locking mechanisms which allow them to connect to the

5-3 Bull float with adapter and extension poles.

bull float adapter and also lock into each other. For a slab measuring 18 feet, you would connect one 6 foot pole to the bull float and then two more in succession to the first pole for an overall reaching span of 18 feet.

A bull float is pushed across and then pulled back over wet concrete slabs to remove tamping marks and generally smooth surfaces to an even and glassy texture. This tool is also used from various angles to fill in cat's eyes—small and shallow low spots easily noticed on slabs that the bull float never touched. Maneuvered from different angles, bull floats can pick up cream from surrounding areas to help fill in shallow cat's eyes.

The swivel feature on a bull float's extension pole is used to position extensions in such a way that the float can be pushed out onto a slab surface with its front edge lifted slightly. This prevents the tool from digging into the concrete.

FRESNO

The fresno is a finishing tool that looks and operates much like a bull float. The only difference is that instead of an operating surface made of wood, a fresno features a metal blade (FIG. 5-4). Its purpose is to continue smoothing concrete as it stiffens and becomes harder. The extension poles used for a bull float are the same used for a fresno.

5-4 The rectangular fresno on the top with square corners often leaves lines on fresh concrete. Fresnos with round corners are more line-free. Note the adapter on the round fresno, which can be adjusted a number of ways.

Fresno blades are 3 feet long and about 6 to 8 inches wide. They come in two styles; one with a rectangular shape and square corners and another with rounded edges and no square corners. As with a rectangular bull float, square-edged fresno blades tend to leave lines behind after being operated on fresh concrete. The rounded model, on the other hand, tends to be line-free and is a preferred option.

Fresno use is determined by concrete surface stiffness. Used on concrete that is still very wet, a fresno will not accomplish much more than turning a top layer of cream into a soupy mix. Novice concrete finishers tend to work concrete too much. By doing this, they continually agitate the top layer of cream until it becomes saturated with sand and grit. As a general rule of thumb, a concrete slab that has been poured relatively wet should only be bull floated a maximum of two times and smoothed with a fresno about three times. The use of these tools are covered in more detail in chapters 10 and 11.

HAND FLOATS AND TROWELS

A hand float is basically a piece of flat, smooth wood with an attached wooden handle measuring about 16 inches long and 3 inches wide; larger and smaller sizes are available. Hand floats are used to help place, move around, and smooth wet concrete as it is being poured into tight spots, against walls, and along edges. Like a bull float, they are used to flatten concrete while it is still wet and easily workable. After concrete starts to stiffen, wood floats are cleaned and put away, making way for metal fresno tools and metal hand trowels.

Hand trowels are metal blades equipped with handles. They range in size from about 12 inches long and 3 inches wide to 18 inches long and 4 inches wide. Most professionals prefer to use a small "scrubber" hand trowel (14 inches long and 3 1/2 inches wide) to work up a layer of cream on a rather rough surface and then finish it off smooth with a "cheater" trowel that might measure 16 inches long and 4 inches wide. They prefer this method because it is easier to quickly and vigorously operate a smaller trowel for scrubbing and then simply wipe it smooth with a swipe or two from a larger trowel.

The blades on metal hand trowels must be in excellent condition to smooth concrete without leaving behind lines and other imperfections. Nicks and dents along a trowel's edge will scar concrete surfaces and make the finishing effort a difficult one. When renting hand trowels, be sure to inspect the bottom side and all four edges. Should you detect imperfections, ask to rent a different tool; or have the rental yard service department file the edges clean and even.

As a knife blade becomes sharper as it is slid over a honing stone over and over again, continuously operating a metal hand trowel over concrete causes its edges to sharpen too. Never run your hand or finger across the edge of any hand trowel. Like a razor, these edges can quickly slice through skin.

You'll need at least two metal hand trowels, one smaller than the other, to finish almost any concrete slab. Along with working up cream with the small one and actually finishing with the larger (scrubbing and cheating), you'll normally have to lean on one trowel in order to reach spots in the middle of a slab with the other. Unless your concrete job is only a narrow walkway, you'll find yourself leaning on one trowel to reach even further to finish areas away from the edges.

Pool trowels function exactly the same as metal hand trowels except that the front and back ends are rounded. As with a round fresno, pool trowels work much better to finish concrete because they will not leave lines behind. If possible, rent at least one rounded pool trowel to work as a cheater. Should you want to purchase your own, be sure to use the term *pool trowel* because these tools were originally designed to finish the mortar and plaster used to coat in-ground swimming pool sides, which are seldom perfectly flat all around.

Professional concrete finishers usually carry their hand tools in a large, five-gallon plastic bucket. Besides the convenient size for carrying tools to and from work, these buckets are partially filled with water at job sites and used to clean trowels after each use. After rinsing off all concrete, trowels are hung over the edge of the bucket to dry. **CAUTION**: be extremely careful washing metal hand trowels with your hands. Their edges can quickly cut you. Consider using a small brush with rather soft bristles.

KNEE BOARDS

Because the final finish on concrete slabs is achieved by use of metal hand trowels, you are probably wondering how the center part of a slab measuring 14 feet or more is reached. Knee boards are the answer. Knee boards, which generally measure about 1½ to 2-feet square, spread finishers weight over a wide area (FIG. 5-5). Because concrete cannot be hand finished until it has reached a fairly hard texture, concrete finishers rely on these small pieces of smooth plywood to kneel and walk on while maneuvering over large slabs. By the time concrete sets up enough to support finishing, it is also hard enough to support the weight of finishers.

Finishers always use two knee boards. This way, they can leapfrog around slabs by placing one down and standing on it while positioning the other, and so on. Finishers start in a corner where the first mud was poured and then move in a parallel direction to how the slab was

5-5 Knee boards are required for all slabs that are too big to reach from outside forms. This one has a small piece of wood nailed to one side to use as a handle.

screeded. Essentially starting on the concrete that was poured first and ending up in the area where the last concrete was poured.

Knee boards *do* leave impressions on concrete, however. But as long as the wood on their bottom surface is smooth and not pockmarked with knots and gouges, imperfections can be quickly and easily wiped out. A scrubber is used to roughen the area and bring up cream, which is then wiped out with a cheater to a smooth and line-free finish.

If you were to simply walk and kneel on a slab to finish it, your feet would leave deep impressions, knees would dig in, and toes would severely scratch and mar the surface. On that note, be sure your feet rest on the knee board positioned behind you as you kneel on the other board during finishing. Also, because slightly wet concrete can cause knee boards to stick to slab surfaces quite strongly, many finishers nail a small strip of wood to one edge for use as a handle.

Because knee boards leave impressions on concrete surfaces, always try to finish as much concrete as possible from each knee board site. Reach out as far as possible while leaning on one trowel. Finish areas in front of and on both sides of knee boards before moving on. In addition, be certain that knee boards are not spread out too far from each other. Ideally, they should be set in such a way that your knees easily fit on one while your toes can rest on the other. Work your way from front to back or side to side around the slab. Once you reach a form, step off to complete that section; then start a new path. Knock any mud or debris off your feet before stepping on knee boards to prevent that debris from accidentally falling onto the slab surface.

EDGER

Concrete "edges" refer to the very top corner of concrete slab sides; the 90 degree corner produced at those points where a slab surface stops and the actual 4 inch side of the slab begins. Essentially, edges are those bordering points where the top portion of concrete slabs touch forms. By looking at almost any existing concrete sidewalk or patio slab, you should notice that these edges are slightly rounded off, instead of being perfectly formed at a sharp, 90-degree angle.

Rounded concrete edges are purposely made with a hand or walking edger (FIG. 5-6). Creating rounded edges is this tool's sole function. Similar to a metal hand trowel, edgers feature a distinctive, rounded-off side. It is this side that is introduced between tops of forms and concrete. As it is smoothly maneuvered along outer slab perimeters, concrete cream is shaped to conform with the curved arc of the edger tool. Although 3/8-inch arcs seem to be the most popular, these arcs range in size from a tight 1/4 inch to a full 1/2 inch.

Walking edgers are almost identical to hand-held models except that they have long handles. Instead of having to squat or kneel down to use a hand-held edger, walking edgers allow finishers to walk along slab perimeters to effect rounded edges. Walking edgers are also almost always larger than hand-held tools. Although the larger size creates identical arcs, it is

5-6 The walking edger on the left and the hand edger on the right have identical rounded arc sides.

easier to produce crisp edges while finishing a wider swath along slab borders. When either purchasing or renting a combined set of edgers, be certain the arc for the hand-held tool matches the arc for the walking one.

To make the job of concrete finishing a little easier, many professionals like to lightly edge concrete soon after applying a bull float. Although concrete is still much too wet to support a crisp edge, all of the aggregate is pushed down far enough so that future edging maneuvers will not kick up aggregate obstacles. As fresno operations continue, more and more cream is deposited along edges to fill in low spots that might have been produced by early edging. This is good because it provides an excellent surface for edgers to work on later while affecting final edging maneuvers.

SEAMER

Along with expansion joints, concrete finishers install control joints (seams) about every 4 feet along walkways, driveways, and at other crack-hazard points. For the most part, control joints are deep grooves put into concrete finishes in anticipation that if cracks do occur, they will follow seams and not randomly appear anywhere else. Along with expansion joints, seams help to control concrete cracking as much as possible.

Seamer tools are made of metal and feature a raised ridge down the middle of their face (FIG. 5-7). To insert seams, finishers lay a straight 2-×-4 board across a freshly bull-floated slab to rest on forms at both sides of the concrete. The correctly positioned 2×4 is then used as a guide to run a seamer against while forcing it into the concrete. The result should be a definitive line across concrete where rocks and aggregate are pushed down and cream is smoothed.

As with initial early edging, don't expect that first seam to stay perfectly clean and crisp after maneuvering a fresno over it. Cream will surely fill the groove to some degree. Because aggregate has been forced down and replaced with just cream, however, you should have no problem

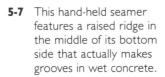

5-7 This hand-held seamer features a raised ridge in the middle of its bottom side that actually makes grooves in wet concrete.

refinishing seams at a later stage. In fact, with a steady hand or use of a walking seamer, you might not even have to use a 2-×-4 guide again.

It is common to place seams every 3 to 4 feet along walkways with an expansion joint every 9 to 12 feet, whichever works best proportionately. Try to maintain evenly spaced seams and expansion joints, not only to prevent uncontrolled cracks, but to look visually attractive. For example, if your walkway will be 24 feet long, plan on installing an expansion joint in the center (at the 12-foot mark) and then put in seams at the 4, 8, 16, and 20-foot marks.

Also, plan on installing expansion joints or a combination expansion joint and seam where walkways or other slabs form a corner so the corner is outlined in a square pattern. An example would be a walkway that comes off a patio slab located at the back corner of your house and then goes down along the side toward your front yard. The point where the walkway meets the slab should have an expansion joint and the point at which a line would designate a 90-degree angle with the expansion joint should have a seam. In essence, the seam would be straight in line with the back of your house to make the first part of the walkway squared off—expansion joint on one side, seam on the other, and lawn on the two open sides.

BROOM

The most common finish for exterior concrete flat work is called a broom finish. It is very easy to put on and hides a multitude of minor hand trowel flaws. It also provides concrete with much-needed traction qualities when wet. Smooth-finished concrete is as slippery as ice when wet. Have you ever noticed how slippery your garage floor is after parking a rain-soaked car on it? Outdoor concrete must be broomed, or finished, in ways other than simple trowel finishing if you expect to be able to walk on it safely during wet weather or any other time the surface is wet—like when hosing it off or while watering the lawn.

Finishers use a soft-bristled push broom or one that is specifically designed for concrete work. Stout-bristled brooms frequently used to sweep asphalt and other coarse surfaces are much too strong and will leave very deep and uneven lines in concrete. Soft push brooms work well to effect attractive lines that not only look good and cover small imperfections, but also cover wide swaths with each pass.

In most cases, the tiny lines left behind as a result of pushing or pulling a soft-bristled push broom across a finished slab scratch the surface just enough to allow adequate traction. If you are pouring a steep driveway and anticipate slippery winter weather conditions, consider using a strong-bristled broom to effect heavy, deep lines to really aid traction.

To reach parts of a slab located further away than your push broom handle will allow, place an extension pole over the broom handle with the featured hole of the extension pole positioned closest to the broom bristles. Through that hole, insert a small wood screw into the broom's wooden handle. This should secure the extension to the broom so that it cannot be pulled off (FIG. 5-8). To that extension, you can attach additional extensions just like you did for the bull float and fresno. If the slab is still too long, or obstacles like fences are in your way, you can operate the broom from opposite sides of the slab to create the broom-finish effect by brooming half the slab from one side and the second half from the other side.

Broom lines do not have to be necessarily straight. Wriggling the broom as you pull it toward you creates a wavy broom design. You could also use a small foxtail broom to create a pattern of small semi-circles. Get back on the slab with your knee boards and simply operate the foxtail

5-8 A soft-bristled push broom has been reinforced with brackets on each side, and a screw in the center of the handle close to the broom head is used to secure extension poles.

broom in "half moon" arcs until you achieve the finish desired. The final design will look something like a windshield wiper pattern with lots of small lines following a series of half circle patterns.

Concrete must be set up and trowel finished before a broom is maneuvered over it. If your concrete has set up so much that a soft-bristled broom will not make any lines, try putting pressure on the bristles by pushing down on the broom handle while pushing and pulling it across the slab. If that doesn't work, dip the bristles in a bucket of water so they are soaking wet and then apply them to the surface. Place it as far away as you can and then just pull it toward you. In extreme cases, you will probably have to both push and pull it across the slab.

PROTECTING SURFACES

Concrete work can be very messy. No matter how careful you are, concrete usually gets splattered on something—usually the walls along your house. To prevent unnecessary concrete splatter on the house or other structures, use heavy-duty duct tape to secure newspaper or plastic on walls before pours. Plan to cover an area that reaches from the concrete's top surface to at least 3 feet above. This should protect against splattering from wheelbarrow dumping and tamping maneuvers. Newspapers or plastic are removed while out on slabs with knee boards hand finishing. Helpers may be needed to assist in getting splatter protection away from slab areas. Any concrete residue that falls off onto slabs is finished into surfaces with a hand trowel.

Do not place newspaper or plastic so low that its bottom edge is allowed to touch a concrete surface. Should it become buried in concrete, a slab's finish will be marred when you attempt to pull it out. In addition, should a lower section break loose from a wind gust, its flapping action will make all sorts of marks on any slab surface it touches. If need be, chalk a line an inch or so above proposed slab surfaces to designate the lowest point splatter protection material can be placed.

Besides house walls, there may be other items you need to protect against concrete splatter. Railings, metal patio cover posts, sliding glass door sills and pool pumps are just a few. Heavy-duty duct tape may be wrapped around items or used to secure strips of plastic in place. If concrete does find its way to an unprotected surface, clean it off as soon as possible with a wet rag or sponge. Once it hardens, concrete is difficult to remove and may require use of a wire brush.

RUBBER BOOTS AND GLOVES

The cement powder used to mix concrete is a powerful agent. Not only will it tend to dry out and otherwise ruin normal shoes, it can severely dry out the skin on your hands, resulting in cracks, flaking skin, and other problems. Professional concrete finishers always have a pair of heavy-duty rubber boots available for pouring concrete. Worn during pouring and tamping exercises, they help to keep their feet dry and hold up against the potent effects of cement in concrete. When the work that calls for them to walk around in concrete has been concluded, finishers will usually change from rubber boots into more comfortable leather work boots or other suitable shoes. When the slab has been completely finished and forms pulled, they might put on the boots again while using plenty of water for tool, form, and equipment cleaning.

Plan to wear a pair of heavy-duty rubber gloves while pouring concrete. You will often find yourself reaching into the mud to pull out handfuls of concrete in order to fill tight spots or move it away from the screed board. The lye in concrete dries out the skin on your hands quickly unless they are protected with rubber gloves. Many professionals are forced to wear gloves not only to keep their hands in decent shape, but also to prevent the onset of concrete poisoning. Concrete poisoning is a result of continual exposure to the powerful substances in concrete and causes an infection along hands and arms.

Personal safety during any do-it-yourself project is a major concern. You have chosen to pour your own concrete job to save money and to enjoy the satisfaction of a personal achievement. Do not take any chances on ruining that experience by avoiding the use of such inexpensive safety equipment as rubber boots and gloves. For that matter, plan on wearing eye protection, too. Even plastic sunglasses offer protection against accidental concrete splatter in your eyes. Should this happen, however, thoroughly flush your eyes with plenty of clean water.

Chapter **6**

Site
preparation

Preparing an area for concrete work involves a number of chores. First, remove obstacles from the immediate job site that might hinder overall maneuverability. Patio furniture, swing sets, planter boxes, and the like should all be relocated to a remote part of your yard so that they are totally out of the way. You will need plenty of room to operate long bull float and fresno extensions. Site preparations might also include removing old concrete, relocating sodded lawn sections, transplanting flowers and shrubs, pulling old tree stumps, or digging a desired grade. Systematically plan your site preparation job so that each chore is completed before starting a new one. In other words, don't start haphazardly digging in the middle of a site until you have established the required grade by way of forms, string, and level.

REMOVING OLD CONCRETE

Front doors, sliding glass doors, and other residential door entries often have small concrete steps or stoops at their outdoor entrances. They are usually only big enough to accommodate a small welcome mat and a place where you can remove muddy shoes before entering the home. When planning concrete slabs or walkways in areas surrounding them, plan for their complete removal as opposed to just pouring new concrete around them (FIG. 6-1). Unless the actual size of the steps or stoop fits within planned expansion joint or seam patterns, they will appear awkward and out of place.

Old concrete can be taken out with a jackhammer or busted up with a sledgehammer. For small jobs like stoops, most professionals prefer the latter. With a pick and shovel, remove as much dirt as possible from beneath the outer stoop edges to create an open space under the concrete. Then, force one end of your pick under the stoop to raise its front side off the ground and loosen its grip on the foundation it is positioned

6-1   Rather than pour new slabs around small, existing stoops or landings, remove old concrete so fresh pours look uniform.

6-2   Use a heavy-duty pick or large pry bar to raise the front of old stoops and loosen them from the base. This worker has his foot on a pick's wide blade and is pushing down on it for added leverage.

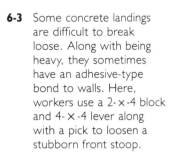
6-3 Some concrete landings are difficult to break loose. Along with being heavy, they sometimes have an adhesive-type bond to walls. Here, workers use a 2-×-4 block and 4-×-4 lever along with a pick to loosen a stubborn front stoop.

against (FIG. 6-2). Use the pick to take advantage of any available leverage. Concrete is very heavy and you might be able to only loosen a little at a time. If need be, establish a better leverage system by using a long, 4-×-4 post as a lever blocked with short 2×4s (FIG. 6-3).

Once the stoop has been broken loose, use the pick to pry it away from the house. Try to slide it straight out horizontally, as upward movement could chip or crack existing plaster or wood siding immediately above it. Moving it out should eliminate any chance of causing damage to the structure while pounding on it with a sledgehammer. Once the old concrete is broken up and removed for disposal (FIG. 6-4), you can begin forming.

FORMING

The quickest and simplest way to start almost any forming job is to lay forms out on the ground in their planned locations (FIG. 6-5). This allows you to better perceive the dimensions of your planned slab or walkway. It also allows you to strategically place each form so that few, if any, have to be cut to fit. In other words, don't waste good lumber. Measure forms and place them in spots where full advantage can be taken of their length instead of cutting good long forms to fit where shorter ones would work fine.

Use forms to their greatest potential. If one side of a slab measures 13 feet, use a 14-foot board and allow it to extend past the corner by a foot

6-4 A tarp protects the front door as concrete is repeatedly smacked with a sledgehammer until a pick is able to pry out concrete chunks in small, manageable pieces.

6-5 Outline proposed jobs with form lumber before pounding stakes. This gives you the opportunity to properly place forms and take full advantage of form sizes without having to unnecessarily cut long forms to fit.

(FIG. 6-6). The extra extension will not cause any problems and you will prevent wasting an otherwise functional 14-foot, 2-×-4 form.

Plan on having scrap lumber on hand that can be cut to size for jobs requiring 1- or 2-foot form boards. If scraps are not available, purchase short 2×4s that you can use to cut out a sufficient number of small forms. Along with preventing waste, preserving long forms means they will be on hand for future concrete flat work projects or to use in another building project.

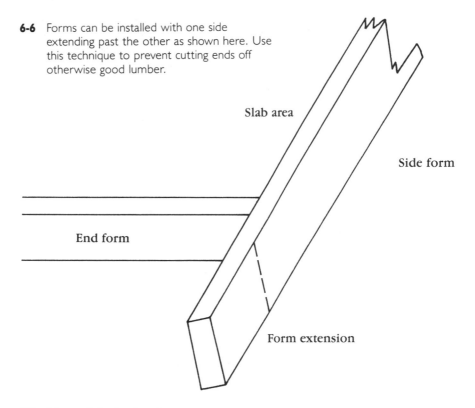

6-6 Forms can be installed with one side extending past the other as shown here. Use this technique to prevent cutting ends off otherwise good lumber.

Slab area

Side form

End form

Form extension

Working with the landscape

In many cases, areas planned for concrete are already landscaped with grass, flower beds, or mulch. Chances are, those areas must be excavated to make room for forms to fit in such a way that their top edges are positioned according to the grade level planned for the concrete. Simply resting on top of a lawn, for example, forms might sit too high in relation to where you want the top surface of concrete to be located.

In this case, lay forms in their relative positions and measure how high they sit with relation to where you want them to. Extending perpendicular from the house, you can measure the distance from the top edge of the form butting against it to where you actually want the top of the form to be. This measurement tells you how much dirt has to be removed in order for the form to fit according to grade. Knowing this, you can accurately judge how much dirt has to be dug out rather than guessing and digging more than necessary.

Further, by having forms laying close to their prescribed positions, you can use them as guides to establish actual digging perimeters (FIG. 6-7), especially when digging out lawns. Using a straight guide helps you to confine excavation efforts to the slab area with minimal disruption to the rest of the landscape. It saves unnecessary digging work, too. Square point shovels are great for this work, as their flat blades cut sod evenly and can scrape out dirt smoothly.

6-7 Once a job is outlined, you'll have a good idea where forms are to be placed. Use a square shovel to cleanly dig out sod or other ground material. A neat job will save extra landscaping renovations later.

Blocking forms together

Slabs and walkways are easiest to form using lumber that has identical dimensions. A 10-×-12-foot slab poured against the back of your house, for example, is most easily formed using one 12-foot and two 10-foot 2×4s. This is great for jobs featuring such measurements, but what about long walkways that are more than 30 feet?

Because locating straight, 2-×-4 form boards in excess of 20 feet is difficult and expensive, you will have much better luck blocking two shorter 2×4s together to make one long straight form. For a 30-foot run, a blocked form consisting of 16- and 14-foot boards is perfect.

Lay the boards down end-to-end on a flat surface. Determine which side of each board you want facing the concrete (their best sides) and which edges you want facing up (their best edge). Put the good side face down and note which edge is face up by marking it with a pencil. This will guarantee that after the boards are blocked together, their best faces will be located against the concrete and their best edges will be on top to make concrete edging easiest.

Drive a nail part way into the far end of each board. Attach a string to one nail and then stretch it tight to the other nail. Secure it with a good knot. This string will serve as a guide to show that both 2-×-4 boards are aligned with each other. In lieu of a string, you could butt another long 2×4 against the two on the ground to ensure they are lined up correctly.

Use a 2 to 4 foot, 2×4 as a block. Center it over the spot where the two 2-×-4 forms butt together. Securely nail it to both forms using at least six to eight nails on each form (FIG. 6-8). While nailing, constantly check your string or 2-×-4 guide to ensure the forms remain in a straight line. This is easiest by completely nailing the block to one form first and then the other. Be sure the block's top edge matches the form lumber's top edge. If it sticks up higher, it could throw off screeding levels. In addition, have someone help you carry blocked forms to their working location to prevent bowing or weakening the block.

6-8 Two long 2-×-4 boards butt next to each other to form part of a driveway pour. A short 2-×-4 block has been nailed to both forms to strengthen the joint and keep it straight.

ESTABLISHING THE HIGH POINT

The high point of any concrete job determines the direction that water will drain. Typically, high points are almost always located next to house foundations or other walls so water will run away from them and toward a yard, drain, or street. Deciding on the correct high point of any slab or walkway is critical. Misplacing high points can cause water runoff to flow toward a house or structure and puddle, eventually causing dry rot, mildew, and other water-related problems.

When pouring concrete next to front or rear doors, use an existing doorsill as a guide (FIG. 6-9). (Always attempt to remain at least 2 inches below floor levels to prevent water, dust, and debris from easily entering through door spaces.) Make pencil marks on the wall at each end of the sill, level with its lowest edge. Measure down from those marks to where you want the top of the concrete to be (FIG. 6-10). Join these two marks

6-9   A carpenter's level is used to designate where a pencil mark should be made to the side of a doorsill. This mark will be used as a high-point guide.

6-10   Pencil marks located an equal distance from each side of the door show where forms should be placed so that the door is centered in the middle of a new concrete slab. Pencil marks are also used to determine form height against the house.

together with a chalk line. Chalk lines are also used to denote grade levels along the sides of structures, such as down the side of a house to show where a walkway surface will be or along the back wall to designate the top of a patio slab (FIG. 6-11).

Before chalking any line, though, grade levels have to be determined at each end and then marked with a pencil. These marks are measured down from a wall's plaster lip, the bottom of a horizontal piece of siding, or any other predominant feature that is installed level.

For a walkway down the side of a house, for example, the mark at the back corner of the house could be located 2 inches down from the wall siding's bottom edge, while the mark at the front corner of the house will

6-11 This chalk line denotes where a slab's top surface should be. This helps while grading and during concrete placement.

be located 4 or more inches down from that siding's edge. A chalk line snapped between these two marks will show that the walkway will run downhill from the back of the house toward the front. Note: snapping long chalk lines might require three people; one at each end to hold the line on their marks and one person in the middle to actually pull the line away from the wall and snap it (FIG. 6-12).

6-12 Snapping long chalk lines against a wall or foundation might require three people. Two on the ends to hold string tight, and one in the middle to snap the line.

SQUARING FORMS

The corners of rectangular slabs and walkways should be formed at right, 90-degree angles. To ensure that forms start out square when pouring patio slabs and other flat work against a house or other structure wall, use a carpenter's square. Position one side of the tool along the affected wall of the structure and adjust the form until it sits perfectly square with the other arm of the tool. Secure that form with stakes and set the next form square with it. Continue around your slab making sure all corners are square before actually securing them to stakes.

Squaring forms for concrete jobs where one side abuts an existing structure, like patios and walkways next to houses, is easy because you have something straight to work off of. Setting up a square slab evenly distanced away from structures entails a little more work and ingenuity. For this type of form squaring job, you'll need a roll of sturdy string line and at least six stakes, along with a carpenter's square. This job is easiest when two people work on it together.

If one side of your new slab is designed to be in line with an exterior wall of your house, place a stake right next to that wall so string can be stretched from it along the wall for about 10 to 12 feet. Now, grab the roll of string and stretch a line to about 2 feet past the farthest side of the proposed slab's edge. Move the string back and forth until it lines up perfectly with the house wall. Place a stake in the ground at that point and secure the string to it. This string should be perfectly square with the corner of your house. Use a carpenter's square to be sure that it is, however, and adjust as needed according to the square tool. Place your first form in line with this string and square up the remaining forms to it.

For slabs positioned more or less in the center of a house wall, where side walls cannot be used as just described, you will have to be a bit more inventive. Initially, place a stake along the wall of the house that parallels the proposed slab and denotes the point at which one of the perpendicular sides of the new slab will be located. Attach a string to it and stretch it out to a point about 2 feet past the proposed slab's farthest side.

Have your helper hold one arm of a carpenter's square against the house wall with the other arm next to the string. He can then signal you when the string is square. Place a stake at the square point and secure the string tightly to it. Check the square again to be sure the string is placed correctly. This string will give you a straight line to use as a guide for setting the first form. Use a carpenter's square to correctly position forms that butt against this first form and then again while setting the form parallel to it (FIG. 6-13).

OTHER TASKS

Sometime before actually starting the forming process, you should prepare and plan for other tasks that will require your attention. Although leveling and securing forms is pretty straightforward, setting up a system for screeding might rely on which direction concrete will be delivered from. Screed work always starts at the point where concrete is first

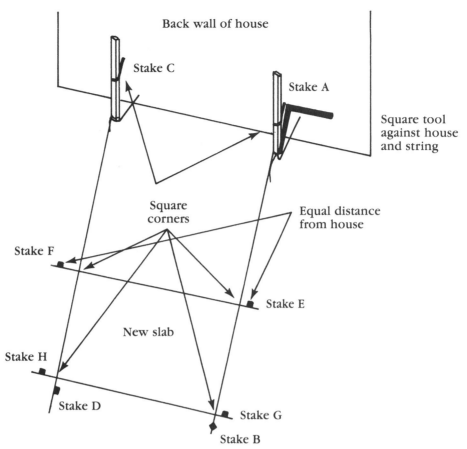

6-13   Forming a perfectly square slab in the middle of a yard parallel with an existing structure without using special equipment can be done with stakes, string, and a large carpenter's square. The distance between stakes A and C denote slab length. Stakes E to G and F to H show slab width. This job starts with stake A and runs alphabetically through H.

placed. Can a concrete truck fit into a space on one side of the slab or will you have to wheelbarrow concrete from the other? Will you need to build special ramps up steps or over other obstacles for wheelbarrows en route from the concrete truck to the work site?

Is there a plumbing cleanout located inside your proposed slab area? If so, you will need to keep it free of concrete so that it can be accessed. Some concrete finishers cover cleanouts with large coffee cans placed upside down over them with their bottom sides evenly lined up with the slab surface. When access to the cleanout is needed, the bottom of the coffee can is simply cut out. These items can also be protected with ordinary lawn sprinkler control boxes, which are quite similar to covers used to protect residential water meters.

Horizontal vents located along house foundations must be left clear.

To hold concrete back from them, use 1-×-4 form boards. Are basement window levels located below your proposed slab surface? You will not want to pour concrete against them, so plan form work accordingly.

Will you eventually want to extend a water or electrical line from your house to an area beyond the new slab or walkway? Provide for that possibility by inserting conduit or plastic pipe under formed areas so that the option is always open.

Discarding old concrete and landscape debris might be a problem, especially if you do not own a pickup truck. You will have to arrange a means of disposing of unwanted debris.

Finally, as you prepare to actually start forming, lay out your job in such a way that you will not have to continuously get up and down to retrieve stakes, nails, tools, and the like. With forms placed as designed, lay down a stake every 3 feet or so around the entire perimeter. Use a nail bag filled with plenty of duplex nails and the tools you will be using. This way, once you get a form correctly positioned, you can grab a nearby stake, pound it in, and nail it with a minimum of effort.

Chapter **7**

Basic forming techniques

*C*oncrete forms are strategically placed and secured to accomplish one goal—to keep wet concrete confined to a certain space in a predetermined shape until it hardens and can remain stable on its own. This is a simple concept. However, should these forms be placed in an illogical manner without regard to surface height, straightness, or grade, the end result will be a hardened slab of concrete that looks awkward and unprofessional and might very well prove to be functionally inadequate.

Forms must be strongly braced with plenty of stakes to ensure they remain in position while being subjected to the weight and pressure of pouring concrete. In addition, every form installation job must account for water runoff, whether it is to assist in yard drainage or to prevent puddles next to a house or other wall structures.

From a visual perspective, forms must be straight in order to flow with the basic symmetry of any landscape. If forms are not installed straight, there is no way concrete will harden straight. Crooked concrete edges wobbling along the sides of slabs and walkways are signs of unprofessional workmanship and cannot add distinction to any landscape design.

FIRST FORM

Most residential concrete patio slabs and similar jobs are poured next to houses or garages, therefore, the following examples depict common designs in which one wall of a house or other structure serves as one side of the slab's forming needs. In other words, concrete will be poured against the bottom of the wall to make it, in essence, one of four required forms. Therefore, the first form you install will actually be a side form.

First, mark the structure wall to designate slab height, as described earlier—at least 2 inches below the bottom of a door sill, house siding board, wall plaster lip, etc. Be certain it is marked at the spot you intend to place your first side form.

Butt the end of your form to the wall so that its top edge is located on the mark. The form should stick out perpendicularly from the wall at a right angle. Check it with a carpenter's square. If the grade is too deep for the form's top edge to meet that mark, push a little dirt under the form to raise its edge. If the grade is too high to allow the form's top edge to go low enough to meet the mark, use a square point shovel to scrape away dirt. Adjust the grade along this form's length until it rests in a position where its top edge is in line with the mark.

Once this form has been closely adjusted to grade and angle, position yourself inside the forming area and pound a stake into the ground right next to the form's outer face about 1 foot away from the house wall (FIG. 7-1). Brace your knee or foot next to the form's inside face while lightly pulling the top of the stake toward you. By sandwiching forms between your foot and the stakes, you are better able to control and guide stakes as you pound them into the ground. Putting stakes into the ground correctly takes practice. They should be straight up and down in order to secure forms in a vertically upright position.

Lay the form against the stake and check it for square. With the form's top edge at the designated mark, or slightly higher, brace your foot against its inside face and drive a duplex nail into the outer face of the stake until it goes through, and into, the form, securing the form to the stake.

Check the square again. If it is off, move the far end of the form horizontally one way or the other until square is established. Once that is done, pound a stake next to the outer face of the form about 1 foot in

7-1 The first stake placed along your side form should be about a foot from the end of the board. Brace your foot or leg against the form to keep it steady.

7-2 Once the form has been positioned, drive a second stake into the ground about a foot in from the end of the form.

from the far end (FIG. 7-2). Use the same foot bracing technique as you did for the first stake. This method is essentially used for placing almost all stakes. Do not nail this stake yet, as the form's proper level must be established first.

Place your level on top of, and in the middle of, the form (FIG. 7-3). Because this side form will run away from the house, the end next to the house should sit higher than the far end. This is designated on the level tool when the bubble is located past the indicator mark closest to the house. Bubbles always rise toward the high side (FIG. 7-4). If necessary, use a square-point shovel to add fill dirt or cut down high spots until the form sits in a slight, downhill run away from the house; all the while making sure the top edge butted next to the house remains at, or slightly higher, than its designated mark.

As a general rule of thumb, outdoor patio slabs, and the like, should fall about 1/4 inch per foot to ensure adequate water runoff capability. If your carpenter's level has two indicator marks on each side of the bubble, plan to position forms so that the bubble touches the second line on the high side. For just single indicator marks, let bubbles rise about an eighth of their length past the mark to establish downhill runs. Once your level indicates appropriate slope, nail the second stake to the form.

7-3 Before nailing forms to stakes, put a carpenter's level on top of forms and check slope. The bubble inside your carpenter's level will rise toward the high side.

7-4 The bubble goes past the left side mark to show that the form is higher on that side. A good slope is when about one-eighth of the bubble is past its mark on the designated high-point side.

7-5 An easy way to set forms at their precise height is to nail them to placed stakes while they sit too high. Once nailed, lightly tap stakes down until forms reach their appropriate height.

Now, if you allowed the form to ride a little higher than the mark on the house, don't worry. You can always pound stakes down a little to force forms down (FIG. 7-5). The nail securing the form and the stake together allows both to move down in unison. If form ends are too low, however, it is not advisable to pull stakes up. In this case, simply pull nails out of the stakes and reposition forms for new nails.

Your next job is to place stakes along the middle of the form at intervals of no more than 3 feet. To ensure form straightness, place a small nail at each end of the form as close to the inside, top corner as possible; then tie a string to one end. Pull the string tight and secure it to the other end (FIG. 7-6). Positioned along the top, inside edge of the form, this string will serve as a perfectly straight guide (FIG. 7-7). As you position stakes along the outside face of the form, push or pull the form until its top, inside edge is in line with, and runs flush to, the string. Pound stakes into the ground as required. Check and adjust the level before driving nails.

Setting stakes can be tricky work. You might have to hold both the form and stake in one hand to push or pull them into position while at the same time pound on the stake to secure it. In rocky soils, you might need to use steel stakes in order to achieve minimum stake depth for form support. Soft soils require long stakes and short stakes are fine for most hard-pan surfaces. Should an end be forced out of square while staking, you can use a kicker stake to force it back where it belongs. Kicker stakes are inserted at an angle from the top edge of forms toward the outside (FIG. 7-8). These stake faces will only ride against the top outside corner of forms to force their top edge in toward slab areas (FIG. 7-9).

Should you be required by the building department to place a footing under the perimeter of your slab, you'll need to use long stakes, which will be difficult to place (FIG. 7-10). Many professionals have discovered it is easier to form jobs first with long stakes and then dig out footings once

7-6 A string is secured to each end of the form to serve as a straightedge. It is located right along the form's top, inside corner as pointed out by the pencil.

7-7 The string guide helps the form installer position stakes in such a way that they keep the form straight. It might have been easier to form this job first and then dig the footing.

7-8 Kickers add support to form tops and help keep them straight. Kickers are extra important on this job because regular stakes have little ground to support them due to the footing.

7-9 Stakes laid across the footing help support the form. A number of kicker stakes are spread out in preparation.

7-10 A form installer braces a long stake at the top with his hand while using his foot to brace the bottom. A string guide is in place and is used as a straightedge.

7-11   If forms must be placed off the ground to establish slab height, a second form can be placed directly under a top form to fill the void.

7-12   Be alert to the guide strings position when placing stakes. Caught on a sliver or other obstacle, it will be off center. Footings are easiest to dig after forms are installed, especially in softer soils. Use a square-point shovel to square footing sides.

all the forms are secured in position. With forms in place, they are able to accurately determine footing perimeters and can dig perfect footings straight down from inside form faces using a square-point shovel (FIG. 7-11). This eliminates the need to bridge forms in place while attempting to insert stakes, a process hindered by the collapse of footing walls (FIG. 7-12).

SECOND FORM

The second form of a slab poured against a house or other structure's wall is the one that runs parallel to the first. One end of it will butt against the house and run slightly downhill away from it. The end form is saved for last. That is the form that parallels the house and designates how far the slab will extend out from it.

This second form will be located a predetermined number of feet away from the first form to provide your slab's length; the distance in which the slab will run along the wall of the house, as opposed to out away from it. The easiest way to pinpoint the location for placement of the second form's near end is to measure across the wall from the end of the first form. Remember, when you attach your tape measure to the outside face of the first form in order to stretch it across the wall, the width of your 2-×-4 form is $1^{1}/2$ inches. Therefore, if the length of your slab is 20 feet, measure 20 feet $1^{1}/2$ inches. The compensating inch and a half allows for the width of the first form. Your mark should then be exactly 20 feet from the inside face of the first form to the inside face of the second form.

At the spot where the second form butts next to the house wall, there should be two marks—one to designate the level (top edge of the form) and one to designate the length of the slab (inside form face). Using your marks as described for the first form, set the first stake, check for square, level, then nail (FIG. 7-13).

7-13 Second side forms must be positioned square with the buildings they come off of and parallel to first side forms. Use a tape measure, carpenter's square, and string to assist in form placement.

You can set the end stake for this second side form in three different ways: with a carpenter's square, as you did for the first form; by using a tape measure stretched from the far end of the first form to a length equal to that along the house wall; or by laying the end form in position, with either an actual pencil mark denoting the proper length (space) between side forms or with it being cut to exactly the correct length. Again, you must check for square and level and adjust the form accordingly. Once you are assured that this second side form is positioned correctly, with stakes at each end, place middle stakes at least every 3 feet (FIG. 7-14).

7-14 Brace your foot against forms when driving nails through stakes and into them. If nails are long enough to penetrate through a form's inside face, be sure your foot is off to the side of the nail's position.

END FORM

As long as both side forms are installed correctly, the end form should be easy to position. If this form is cut to size, you can simply measure down each side form and mark the point where you want the end form to be. On a slab measuring 10×20 feet, for example, you would make pencil marks on the top of both side forms exactly 10 feet from the house wall. The ends of the end form would then be nailed to the far ends of each side form at the prescribed 10-foot pencil marks. As long as both side forms are set to grade (proper slope level), and you nail the end form so that its top side matches the side form's top sides, its level should be correct (FIG. 7-15).

A word of caution; just because lumber is supposed to be in even-

7-15 Nails are driven from the outer face of a side form and into the end of an end form. Hold end forms in position while nailing.

foot increments does not always mean boards are cut exactly to length. A 10-foot-long board, for instance, could easily measure 10 feet, 2 inches, or any such length just over what you would expect. Therefore, it is always a good idea to measure form boards before using them.

With both ends of the end form securely nailed to the side form ends, place stakes as you did for the other two forms. Be sure to check for level as you go (FIG. 7-16). Use a string to ensure the form is staked straight. To double check corners, use a carpenter's square (FIG. 7-17).

In lieu of relying on a string guide to help stake the end form straight, you might be able to use a block. In this case, a block would be a 2-×-4 board that measures the exact distance you want the slab to extend out from the wall; on a 10-×-20-foot slab, it would be a 2×4 that measures exactly 10 feet. Lay this board on its narrow side with one end against the inside face of the end form at the point where you want to insert a stake. Place the other end of the 2-×-4 block against the house wall and check square.

At this point, stand on the outside of the form, and place the stake against the end form with your foot braced against it, essentially wedging the stake between your foot and the form with the block holding the form steady. This should allow you to quickly stake the end form with its inside face exactly 10 feet away from the house wall. Check the level and then nail the stakes to the form. Using blocks to support forms while installing stakes is a great time-saver, especially for patio slabs, narrow jobs, and walkway projects.

7-16 A carpenter's level rests on an end form as a stake is pounded into the ground. Before this form is nailed to the stake, it is checked for proper slope.

7-17 A small carpenter's square is used to check a 90-degree corner while setting a form. The top form has already been secured.

KICKER STAKES

Kicker stakes were mentioned earlier in the chapter as a means of forcing top form edges in toward slab areas when they have been accidentally forced out while installing other stakes (FIG. 7-18). Not only can kickers adjust form edges, they also add a great deal of support to form tops. This

7-18 Kicker stakes are driven in at sharp angles along top, outer form corners.

is especially important for pours with extra long runs where you want to be sure top concrete edges are as straight as possible, such as patio slabs that run the entire length of a house or a long driveway (FIG. 7-19).

Kickers are an absolute must for jobs that require footings. This is because upright stakes might only have an inch or two of dirt between them and the open footing (FIG. 7-20). Because of the enormous weight of concrete, all footings are completely filled with concrete to the base of forms (grade level) first, before any concrete is placed against forms to create an actual slab. This gives footing concrete time to settle and actually lend support to upright stakes by filling the footing void.

GRADING

Once forms are secure, the outline of your concrete job will be clear. At this time, you must concentrate efforts on making the entire area inside forms conform to an even depth of no more than 4 inches. This goal is referred to as *grading*. Endeavors to flatten slab areas to an even depth of $3^1/2$ to 4 inches accomplishes two things: precise dimensions for calculating concrete yardage, and an evenly dispersed concrete depth to help avoid uneven pockets that could lead to eventual cracks. Haphazardly graded slab areas might result in some spaces being filled with 5 to 6 inches of concrete while others only get 2 to 3 inches. Uneven depth causes weak spots that generally crack.

7-19 Plenty of kicker stakes have been placed along this driveway form. The form is solidly secured in position and should not budge during concrete pouring.

7-20 Kickers are required for forms along footing installations to keep forms from "blowing out," or moving out of plumb.

Grading work is most labor-intensive when dirt has to be removed. You might be able to simply scrape soft soil from the surface with a square point shovel, if you're lucky. But more often than not, you'll need to use a wide-blade pick to chop dirt loose, which is then scooped up with a square-point shovel. The grading tips in this section can help minimize the workload of grading and avoid digging more than necessary. These tips help you to set up a grading system that quickly checks excavation depth as you progress across slab areas.

The simplest way to determine grade depth is by placing a screed board on top of forms and measuring the depth below it (FIG. 7-21). The bottom of a screed board determines the top of a concrete slab. Having 3½ to 4 inches open under the bottom screed board edge is great as long as this is a consistent measurement. If you find that the ground you are working on is extra hard, try using the pick end to break up dirt and the wide blade to clean areas out. Do not get carried away and dig deep holes. Rather, rely on short pick strokes applied at an angle to chip dirt out, not hefty blows directed straight down.

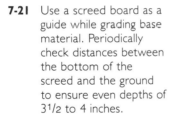

7-21 Use a screed board as a guide while grading base material. Periodically check distances between the bottom of the screed and the ground to ensure even depths of 3½ to 4 inches.

When dirt has been broken up and loosened, you might find that a heavy-duty rake works well to move material out of the way and into piles for later pick up with a shovel (FIG. 7-22). Rakes are good leveling tools and can move a lot more dirt than you might think. In addition, your rake might measure close to 3½ to 4 inches and be a perfect tool for grading

7-22 A heavy-duty rake works well for grading bases.

under screed boards to fine-tune depths. Operating a proper-sized rake along screed boards can give you quick and constant depth reference, as every time the rake fits under the screed board clearly, you'll know that the grade depth is close to 4 inches.

Another way to help determine grade inside formed slab areas is by attaching a grade screed to the bottom of your regular screed board (FIG. 7-23). To make this guide, simply nail two wooden stakes to a 2-×-4 grade screed that measures a few inches shorter than your regular screed board. Make the tops of the stakes flush with the bottom of the grade screed so that when it is scraped against the ground, the stakes won't dig in. Then, simply butt the grade screed to the bottom of the regular screed board and nail the stakes to the regular screed.

In essence, you'll have a screed board resting on top of the forms with a 2-×-4 extension attached to the bottom of it and secured by wooden stakes nailed to both of them. This shorter-grade screed should fit between forms, and as long as the regular screed board remains in contact with forms and the grade screed barely scrapes the ground surface, you'll be assured of grading to a 3^1/2- to 4-inch level.

Another way to gauge grade depth is with string. With nails positioned along side forms, you can stretch string across slab widths and periodically measure from the ground up to string locations (FIG. 7-24). In lieu of constantly using a tape measure to gauge grade depth, make a dark pencil line on a stake 4 inches from an end and use it as a measuring

7-23 This upside down stake is nailed to a regular screed board that rests on top of a form. It is also nailed to a shorter, 2-×-4 board directly under the screed. The bottom 2×4 is a grade screed that accurately helps to denote proper grade depths.

7-24 A string tightly stretched across a formed area shows where the top of a slab is designed to be and assists grading efforts. It was set up with a carpenter's string level and is supported in the middle by a stake to prevent it from bowing.

7-25 A dark pencil mark 4 inches from the end of this piece of wood is used as a quick and handy depth gauge.

device (FIG. 7-25). This is not always the easiest method to use. String can get in the way of excavation work, and it is not nearly as maneuverable as a grade screed.

Sometimes, grade levels are much lower than expected. In this case, you must use fill material. Suppose, for example, you want to pour a patio and you want the top of the concrete to be 2 inches below your sliding glass door sill but the existing grade level is 8 inches below the door. The easiest way to prepare this site and still maintain just a 4-inch slab depth is to form it with 2-×-6 lumber and fill in the area with 2 inches of sand. Optimally, you should leave a 4- to 6-inch-wide trough along the 2-×-6 forms where the concrete will extend from the bottom of forms to the top. This trough will allow concrete to be about 6 inches deep along forms. This way, once the slab is poured and the 2-×-6 forms removed, the concrete sides will extend to the ground instead of exposing a 2-inch base of sand, which could erode.

Sand is an excellent base material for concrete. It grades easily, fills low spots quickly and does not shrink or compress once concrete is poured. In lieu of fill sand alone, you could use pea gravel or a mixture of pea gravel and sand to give your slab a solid and uniform grade base.

SETTING SCREEDS AND SCREED FORMS

Once concrete is on the ground, you must screed it to flatten and level its surface. This phase of any concrete pour is very important. If not done correctly, slab surfaces could be pockmarked with low spots that invite puddles of water and are riddled with waves of high spots that cause benches and tables to wobble, people to trip, and other unnecessary problems. Most small- to medium-sized flat work slabs and walkways enable screed boards to rest directly on top of forms as they are pulled down to level concrete (FIG. 7-26). For larger slabs—longer than 14 feet—

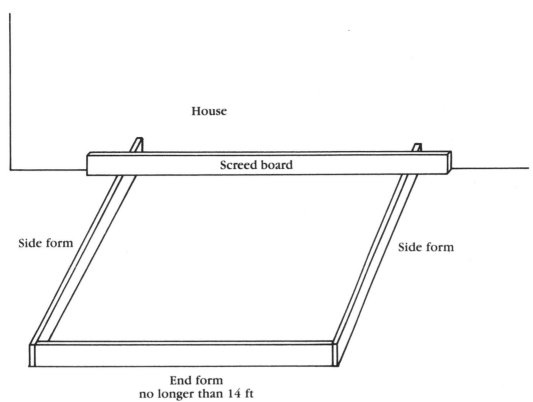

House

Screed board

Side form

Side form

End form
no longer than 14 ft

7-26 Ideally, screed boards are operated on top of side forms as shown.

professionals prefer to install screed forms on which one end of a screed board rests while the other end sits on an actual form (FIG. 7-27).

Screed forms are 2-×-4 boards positioned in such a way that screed boards can be operated on them instead of on top of actual form lumber. They are used for concrete jobs where the spacing of actual form boards is too long or too wide to adequately support a screed board. Generally, if this spacing is more than 14 feet, a 2-×-4 screed board will tend to bow in the middle, causing concrete surfaces to incur a slightly rounded indentation that conforms to the bowed, 2-×-4 screed board.

Although you could install screed forms inside slab areas to the same level as side and end forms, it is easier overall to place them above. This is so you will not have to pull them out of wet concrete after screeding and then fill in the void made by their removal.

Placing screed forms above slab areas means that their bottom side will rest on a form's top side. The screed form will be located at the same level the screed board is operated (FIG. 7-28). In order for screed boards to be supported by a screed form in this manner, a wood stake is nailed to the end of the screed board and allowed to stick out 3 to 4 inches (FIG. 7-29). It is this ear that will rest on the screed form while the screed board

7-27 A 2-×-4 screed form end rests on an existing concrete garage apron while supported in the middle and on the other end by screed pins.

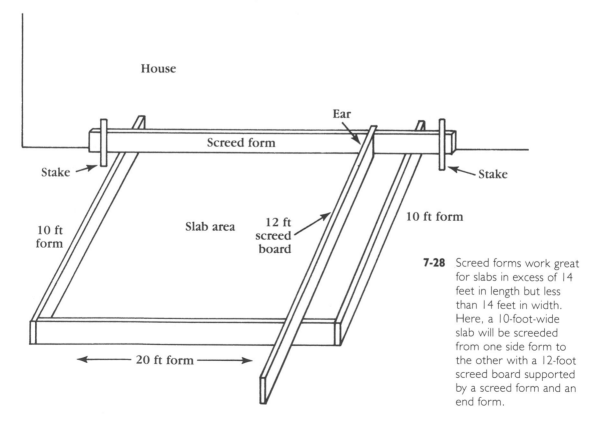

7-28 Screed forms work great for slabs in excess of 14 feet in length but less than 14 feet in width. Here, a 10-foot-wide slab will be screeded from one side form to the other with a 12-foot screed board supported by a screed form and an end form.

is pulled across the slab. The other end of the screed board generally rests on an existing form board. Ears must extend past the end of screed boards by at least 3 inches to allow enough working room for the board to be "see-sawed" back and forth across a slab.

Screed forms are also commonly used on jobs where no regular forms are available to support a screed board (FIG. 7-30). Wide porches frequently present this type of problem. In these situations, screed forms

7-29 To operate screed boards on screed forms, an ear must be secured to an end of the screed board. This system of screeding concrete works well for spans of 14 feet and less. Be alert that screed board ears remain in position during the screed operation.

7-30 Screed forms set up on each side of this flagstone and rock porch will support a screed board with ears on both ends of it. Concrete will be placed next to the porch first and gradually filled in as screeding continues toward the form located at the bottom of the picture.

can be installed perpendicular to walls and other obstacles and supported on their open ends by sturdy stakes. When two screed forms are used to support a screed board, ears must be installed on both ends of the screed board.

Screed forms can also be set up to run parallel with an end form for special needs (FIG. 7-31). Each concrete pour is a little different from others and you might have to devise a unique screed form/board combination. Be aware, however, that anything placed inside a slab area that doesn't belong there, will have to be pulled up after concrete is placed. This could become quite a project if you have a number of screed forms buried at any one time. It is much easier to place screed forms above slabs so that the only thing you have to pull out are stakes. Stake holes are quickly filled with partial shovel loads of concrete and quality tamping makes their impressions in wet concrete disappear.

Installing screed forms entails the same basic work you completed on regular forms. The only difference is that the bottom side of any screed

7-31   The long screed form in front facilitates screeding in an area toward the bottom of this picture. The short screed form at the top will be used just for that concrete section located between the two screed forms.

form will determine the concrete's surface level. A carpenter's level is needed for screed form placement. Use it on top of the screed form to help you set the "free" end when situations do not allow the form to span across two regular forms. The free end will be supported by a stake and the other end will rest on a form.

Once a stake has been pounded into the ground, lay a screed form against it about an inch or two too high. Once it is nailed, put a carpenter's level on the screed form. Gently pound the stake down until the screed form reaches the desired level. Screed forms should be supported by stakes at 4-foot intervals to ensure they are not pushed down while operating a screed board on top of them.

7-32 The arms that extend out from screed pins are adjustable. They have to be in order to meet grade requirements in all soils.

7-33 This screed pin was driven into the ground securely and then its arm adjusted a bit higher than grade as determined by the string guide. With a screed form in place, the pin is hammered down until the bottom of the screed form perfectly matches the string level.

Professionals use screed pins to support screed forms in the middle of slabs (FIG. 7-32). Round steel stakes support metal arms that measure about 1 1/2 inches wide—perfect for holding a 2-×-4 screed form in place. Once screed pins are driven into the ground and a screed form is placed on them, installers can put a level on the form and pound screed pins down until the proper screed form level is achieved. String can also be stretched across a slab area to denote grade. Again, with a screed pin and form positioned high, the pin can be hammered down until the screed form meets flush with the string (FIG. 7-33). Be certain the string is taut; loose string will bow in the middle and sit too low.

By taking a few minutes to visualize your concrete pour in action, you should be able to determine from which direction concrete will come and how it can be properly screeded. Remember that concrete is heavy and screeding is a labor-intensive chore. Long screed boards will have to move and flatten a lot of concrete. Although you should be able to screed spans up to 14 feet with just one screed board, three workers will be needed—two for the ends and one in the middle. If a screed form was set up in the middle of such a pour, you could screed half the area with a shorter screed board that only requires two operators and much less physical effort.

Chapter **8**

Forming walkways

Concrete walkways are formed with 2×4 lumber, much the same as larger slabs. About the only difference is the drastic change in design dimensions. Walkways are generally long and narrow. Although forming slabs featuring narrow widths is relatively easy, maintaining perfectly straight sides along the entire length of long walkways can be troublesome. Be prepared to use string as a guide or plan to use blocks while forming. Walkway forming work requires an adequate amount of time be spent on overall design and installation planning. Will you want concrete poured right next to your house, or will a planter between the house and walkway be more visually attractive? Will you need to install water or electrical conduit under the walkway for a future landscape sprinkler or lighting system? How about rain gutter downspouts? Will they require drains? And, consider the walkway's width. Most of the larger concrete tools you'll use will be 3 feet long, like the tamp, bull float, and fresno. Will finishing work be made easier if the walkway was formed at 3 feet, 1 inch instead of exactly 3 feet or less?

About the easiest way to ensure board straightness while staking and securing walkway forms is to use blocks. These are 2-×-4 boards cut to the precise width of your walkway (3 feet, 1 inch) and inserted between a featured wall and the forms (FIG. 8-1). With one end of a block butted next to the house or other wall structure, you can lay forms against the block's other end to guarantee that the form is placed exactly the right distance from the wall as planned. Stakes are also placed next to forms in line with blocks. With your foot braced against a stake, sandwiching a form between it and the block, you can pound the stake straight into the ground easily and at the same time keep the form correctly positioned.

Because the same block, or a set of matched blocks, are used throughout a walkway forming job, forms will always be positioned away from walls at an equal distance and also be straight. About the only problem encountered with this system is uneven structure walls, such as those

8-1   A 2-x-4 block is braced against a wall behind the form installer and is used as a brace for the 2-x-4 form in front of it. Blocks help form placing and staking jobs go faster because predetermined width dimensions are continually maintained by rigid and unison block support.

covered with plaster or stucco. On these, you will have to use a string guide.

If your walkway is designed with a planter between it and your house, you can still use blocks to aid forming. Simply cut blocks equal to the width of the planter and use them to install the inside form—the one closest to the wall. Then use longer blocks cut to the walkway width to place the outside form. Just butt them up to the staked and secured inside form.

LAYOUT

Walkways are laid out the same as slabs. Forms are placed in their approximate locations to make best use of each board. Use forms to their fullest potential and try not to cut any unless absolutely necessary. Lumber is expensive and many concrete forms can be put to good use after concrete jobs. Completely cleaned with water and a scrub brush, a lot of do-it-yourselfers have used form lumber to build patio covers, sheds, and other outdoor structures.

Many professionals prefer to string walkway jobs before pounding stakes and nailing forms (FIG. 8-2). Stakes are placed in strategic positions at the far ends of a proposed walkway to line up a guide where forms are to be installed. String levels help to accurately determine where to secure

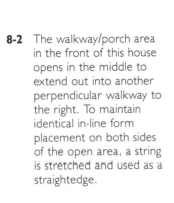

8-2 The walkway/porch area in the front of this house opens in the middle to extend out into another perpendicular walkway to the right. To maintain identical in-line form placement on both sides of the open area, a string is stretched and used as a straightedge.

string on stakes to designate slope. Tightly stretched string serves as a straight-edged guide for forms as well as a slope indicator.

Once all of the height, width, and square adjustments are made with string, form installers can quickly place forms, drive in stakes, and nail the package together accurately. Always be alert to string that hangs up on form slivers, stakes, or nails when working with string guides. Likewise, always be certain string is stretched tight to eliminate bows.

SLOPE

To allow for adequate water runoff, walkways must also be formed and poured with a definite slope running in the direction water should drain. This is a very important consideration. If concrete is poured in such a way that water is able to puddle next to structure walls, moisture can eventually "wick" its way into them, causing dry rot, mildew, and other water problems. At the least, always be sure walkways fall away from walls. This is accomplished by having form ends raised higher along walls than their free ends, which extend out.

Typically, walkways are designed to connect patio slabs in a backyard with driveways or porches in a front yard or just run from the backyard to the front. In most cases, front yards and backyards feature different grade levels. In order for the top of a walkway to meet flush with both a rear patio and a front driveway, it might have to run downhill. This is perfectly fine. However, even though this walkway will slope from front to back, or vice versa, you still need to make certain it also falls away from the house to ensure adequate water runoff.

If your patio and driveway are situated in such a way that both surfaces sit at the same grade level, you can pour a connecting walkway that runs level from front to back yards; as long as slope away from the house is maintained.

The rule of thumb for slope on most outdoor concrete flat work projects is a pitch of 1/4-inch vertical drop per foot of horizontal run. Extending from a house wall, for example, a 3-foot-wide walkway should have a drop of at least 3/4 inch; with the farthest edge from the house wall being 3/4 inch lower than the edge against, or closest to, the wall. This drop must be maintained the entire length of the walkway.

The easiest way to guarantee proper slope is to use a carpenter's level. First, make a pencil mark on the back corner of the house wall to designate where you want the top of concrete to be at that point. Do the same at the front house wall corner. Next, snap a chalk line to connect both pencil marks. Because forms will be laid out and stakes anchored into the ground after using the block system, you can rest one end of a screed board on top of the loose form with the other end matched to the chalk line. Put your carpenter's level on top of the screed board and move its far end up or down until it is level.

Once level, you can measure down from the far end 3/4 inch, 1/4 inch per foot of run on a 3-foot-wide walkway. Adjust the form so its top meets with that 3/4-inch drop designation. It should touch the bottom edge of the screed board. Prop the form up with dirt or rocks to hold it steady. Now, take note of the actual location of the bubble on your carpenter's level with regard to its adjacent line marks. Is the bubble about an eighth of its length past the line closest to the house? At the least, it should read as definitely being ''off-level'' with the high side toward the house. If half or more of the bubble length is past its line mark, the slope is too great. Anything more than a 1/4-bubble length past the mark is too much.

After determining where the bubble on your level will rest while denoting a 1/4-inch slope, you'll be able to lay screed boards on top of forms with the other end flush against the chalk line on the house. Adjust forms to proper height by simply reading the level. You will not have to measure a 3/4-inch drop each time. Do this at the location of each stake so you can nail forms immediately while still keeping your level in sight.

SCREEDS

Unlike a lot of concrete slabs, most walkways are not formed with boards on all four sides (FIG. 8-3). Generally, a house or other structure wall serves as one side. Screeding such walkways can be accomplished in two ways.

Professional concrete finishers frequently wet screed walkways that are poured against walls. They snap a chalk line on walls to designate where the top of concrete should be. Because that line will probably be

8-3 This front porch area requires extensive use of screed forms as a systematic means to accurately screed concrete once it is on the ground. You might have to develop an equally innovative system to achieve acceptable pouring, screeding, and finishing results for your own job.

covered with mud as the pour progresses, they place a second chalk line exactly 3½ inches above it.

Because 2-×-4 boards actually measure 3½ inches wide, the distance between these two chalk lines will be equal to a 2-×-4 screed board. Once concrete is on the ground and hand floated up to the bottom chalk line, a screed board is operated across concrete with its top edge flush with the top chalk line. In essence, the bottom chalk line will designate the correct position for the bottom edge of the screed board and the top chalk line will serve as a guide for the top edge. A keen eye is required for this type of free hand work. Not only must you keep the far end of the screed board on top of the form, but you must be alert to low spots, high spots, and obstructions, all the while maintaining the board's edge on its designated top chalk line.

A more secure way of screeding walkways is by installing screed forms (FIG. 8-4). You can use screed pins or stakes. For extra long walkways, screed pins might be the best option because you can simply slide one long screed form down to the next set of pins after screeding one section of the walk. This eliminates the need to purchase screed forms for the entire length of your walkway. Note: screed pins must be positioned close enough together to fully support a screed form as it is moved from section to section.

Set up screed forms using string guides or chalk lines, whichever is easiest and most versatile. Don't forget that you will have to put an ear on top of your screed board to ride on top of the screed form.

8-4 Screeding maneuvers in this area are best accomplished from the house out to forms. Each screed form support stake is pulled after concrete is poured and screeded, which can be done while tamping.

EXPANSION JOINTS

A combination of expansion joints and control joints (seams) should adorn just about every walkway more than 6 feet long and bordered on either side with another concrete slab. As a rule of thumb, expansion joints should be placed every 10 to 12 feet, divided evenly by the insertion of seams.

A 24-foot walkway, for example, should have an expansion joint in the middle and at each end that is poured against any other concrete slab. Because the middle, in this case, is at 12 feet, seams could be spaced 3 or 4 feet apart to offer a visually symmetrical pattern of both expansion and control joints.

Seams, as you will recall, are grooves made in concrete (FIG. 8-5). For all intents and purposes, they are designed weak spots where concrete should crack first, if it is going to crack at all. A crack along a seam is not nearly as noticeable or unattractive as a crack jogging awkwardly across a concrete surface at various angles. Seams are created by special tools called seamers. As a guide for the seamer, 2-×-4 boards are laid across walkways. Nails driven partially into outer form faces are used to designate predetermined seam locations.

Expansion joint felt is a material specifically made for concrete expansion joints. It is used on city sidewalks, driveways, and curbs. Also, most cities and counties require felt be placed next to public sidewalks, driveways, and curbs whenever new concrete is poured against them.

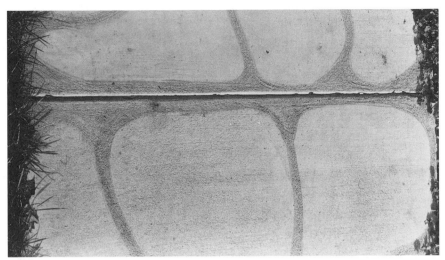

8-5 The deep groove on this walkway is referred to as either a seam or a control joint. It is a weak spot designed to control concrete cracks.

Because the material is not rigid, it has to be supported inside forms with a 1-×-4 brace. That brace is held in place with stakes. Felt braces and stakes must be positioned on the far side of felt so that concrete can be poured next to felt first and then against the brace. This allows concrete to hold felt next to the brace and stakes while mud is poured behind them. Doing it the other way around would cause the brace and felt to be pushed away from stakes.

Once you have placed concrete against felt and about a foot past the brace, you can remove the stakes. Use a hand float to move mud into the holes made by removing the stakes. Then, while slowly pulling up the 1-×-4 brace, force mud under it with a hand float to maintain pressure against felt. Add concrete as needed to fill in voids left by brace removal. If you do it right, plenty of concrete should be positioned against the felt's backside so that it remains in an upright position undisturbed, secure, and straight.

Stringers

In lieu of felt expansion joints, many concrete finishers prefer to insert pressure-treated, 2-×-4 stringers in walkways (FIG. 8-6). Placed inside forms, their ends can be nailed to side forms for easy installation. For walkways poured against walls, one stake is usually sufficient at the wall end while a couple of large nails will hold up to 3-foot-long stringers securely against forms and keep them straight. As with a felt expansion joint, the lone stake is pulled after concrete has surrounded the board.

When using pressure-treated 2×4s as stringers, you must drive nails partly into both side faces. Protruding nails encased in concrete serve as anchors for stringers to prevent them from floating out of concrete during

8-6 All walkway corners must feature either control joints or a combination of control and expansion joints to prevent cracks.

wet weather and from lifting out during cold weather when concrete contracts.

Stringers can be placed anywhere along walkways. Rather than spaced 10 or 12 feet apart just for expansion joints, you can install them every 3 or 4 feet to cover both expansion and control joint needs. The design is up to you and your creative imagination.

Regular 2-×-4 lumber cannot be subjected to outdoor weather year after year. All too soon, it will suffer dry rot and simply break apart. Therefore, always use pressure-treated lumber for stringers. Some finishers have experienced good luck using redwood and cedar, but most prefer material that has been specially treated to stand up under harsh weather conditions. If you are not satisfied with the plain appearance of pressure treated lumber, plan to stain the tops of stringers after concrete has cured. You can stain them a color to match your house, patio, etc.

Benderboard expansion joints

Another alternative to felt material for expansion joints on most walkways is redwood benderboard (FIG. 8-7). Benderboard normally comes in lengths of 10 feet, 4 inches wide and about $3/8$ inch thick. You can install it the same as felt (it is not rigid and must be braced) or split it into narrow, 2-inch strips and put them in wet concrete right after it's tamped and bull floated.

The advantage of using 4-inch benderboard instead of felt is that it is easier to operate a hand edger against. The advantage of using 2-inch benderboard strips is that they do not have to be braced, can be simply inserted into wet concrete, and are easy to hand edge against. But, because of its thinness, 2-inch strips of benderboard any longer than 3 feet are very difficult to keep straight while placing them into wet concrete.

Wet concrete will quickly cover pencil marks, so plan to use nails

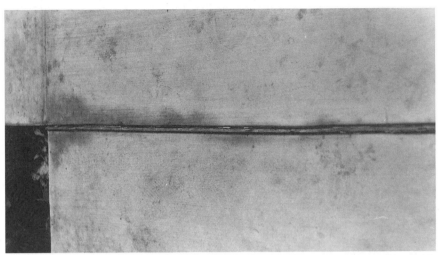

8-7 A piece of redwood benderboard is used as an expansion joint on one side of this walkway corner. Notice the perpendicular seam that squares off the corner. This setup should prevent cracks from occurring inside the walkway corner.

driven into form sides to designate the location of benderboard strip expansion joints. Strips that are cut an inch or so shorter than walkway widths allow a little extra maneuvering room for wriggling them in place while keeping their entire length straight. If need be, gentle hammer taps can be used to persuade strips into position. Lay a board on top of strips and smack that board with a hammer to prevent denting or splitting benderboard. Afterward, use a hand float to smooth concrete around strips.

Should a section of benderboard stick up after concrete has cured, cut it flush to the walkway surface with a razor knife. Sometimes, you can run the blade of a shovel along protruding benderboard strips to split it and make it flush with concrete.

Chapter **9**

Concrete pouring and placement

By far the most physically demanding and time-constraining parts of concrete work is the actual pouring, placing, and screeding of concrete. Although the only way to avoid such labor-intensive work is to hire someone else to do it, there are ways you can do it efficiently, economically, and with a minimal amount of effort. Above all, don't fret time constraints with regard to concrete delivery truck over-time fees. Hurrying a pour without regard to the disposition of screeding maneuvers can cause a tremendous amount of extra work after the truck leaves. And, the amount of money saved will likely be minuscule compared to the efforts of moving concrete around to accomplish adequate screeding.

THE CONCRETE COMPANY

Large, ready-mix concrete companies generally cater to busy contractors who pour lots of concrete every week. Their schedules are tight with lit-tle room to compensate for extraordinary amounts of standby time. Because most of their customers are professional concrete finishers, they can rely on jobs being ready on time and poured within allotted time frames. This is why many large concrete companies might only allow 4 to 5 minutes per yard unloading times.

On the other hand, quite a few smaller concrete companies recognize the inexperience of do-it-yourself concrete finishers and offer unloading time frames up to 10 minutes per yard before any standby fees are incurred. Although they service contractors on a regular basis, their schedules allow for longer unloading periods to better serve smaller resi-dential pours. Likewise, their drivers are more apt to offer advice and recommendations to do-it-yourselfers because they have been involved with so many patio, walkway, and driveway jobs.

The best way to determine which concrete company can meet your needs most economically is to call a number of them on the telephone. Ask about all of their charges and unloading times. Ask the dispatcher if his company makes a lot of deliveries to homeowners or if it mainly serves large contractors. Novice do-it-yourself concrete finishers are recommended to go with concrete companies most familiar with homeowner needs and who have the longest unloading times.

When the concrete delivery truck arrives at your job site, take a few minutes to introduce yourself to the driver. Show him your job layout and explain how you plan to conduct the pour. Mention that you do not have a lot of experience pouring concrete and would certainly appreciate any helpful tips or advice.

Drivers generally have a good idea of what forms, ramps, and other related attributes should look like when properly prepared because they deliver concrete to such a wide range of amateur and professional concrete finishers. They might suggest you install a few extra stakes here and there, provide more solid support for wheelbarrow ramps or may even see a much more efficient and less labor-intensive way to get concrete from their truck to inside your forms. The more congenial you are, the better your chances are of obtaining some very useful advice.

Ready-mix drivers control concrete flow and operate the chute up or down from the back of their trucks. If yours is a job where wheelbarrows are used to deliver concrete to forms, the truck will remain parked and the driver will stand at the back of it to operate controls.

For jobs where concrete will be poured directly from the truck to the forms, drivers will remain in their cabs. From there, they will maneuver the truck and simultaneously operate the chute and control concrete flow according to directions given by a designated person, usually the one maneuvering the chute.

Because the driver must sit in the cab, his vision is limited to side mirrors. Therefore, someone must be within mirror range to provide directions. You could yell and hope the driver hears you, but chances are, the truck's engine noise will limit that means of communication. This is why most professionals use a system of simple hand signals to direct drivers. Be absolutely sure you and your driver have a clear set of hand signals worked out before the pour starts.

Normally, a clenched fist held in the air means to stop the truck from moving forward. This is done when the end of the chute is about $1^{1}/_{2}$ to 2 feet ahead of the last concrete poured. At that point, the chute is swung to the furthest part of the slab and the flow of concrete is started.

If the driver is able to see the job from a side mirror, he will pull the truck ahead as you finish filling a section with concrete. If the job is out of view, you will need to wave an open hand under your chin to designate that you want the flow of concrete shut off. Do this a couple of feet before sections are filled, as concrete will continue to flow down the chute even though the truck's container drum has stopped moving. Extra concrete poured into filled areas will have to be moved out with shovels, a lot of heavy work.

The chute is raised and lowered by a hydraulic ram. During a pour, keep the end of the chute about 1 to 1^1/$_2$ feet off of the ground. This minimizes concrete splatter and helps the chute operator better judge how much concrete is needed to fill certain areas. To have a driver adjust the chute, point to it and then simply point up or down.

Any set of signals can be quickly worked out with just a little discussion. Keep it simple. Should problems arise, stop everything and verbally communicate. Work together to help your job flow smoothly.

FINAL CHECK

Before the concrete truck arrives, walk the job site to make sure everything is ready to go. Put a little pressure on forms with your foot to make sure they are firmly braced and supported. See that conduit pipe, cleanouts, or other obstacles are either covered or secured (FIG. 9-1). Check the location and condition of tools to have them ready at hand but not in the way.

9-1 Make sure forms are secure and other attributes are correctly braced or supported before concrete is delivered. The white pipe sticking out of the black electrical conduit in this formed area is simply taped in place to hold wires out of the way and prevent concrete from falling into the conduit. Note the hog wire on the ground.

Patio slabs and walkways seldom need reinforcing wire or rebar because they won't have to support heavy loads. Driveways are a different story. Because they will have to support the weight of automobiles and delivery trucks, you might consider laying down a layer of 6-×-6-×-10 reinforcing wire, commonly referred to as hog wire. This material features #10 gauge wire fabricated into 6-×-6-inch squares. It is available through concrete companies, lumberyards, and other building supply outlets. Rolls are 7 feet wide by 50 to 100 feet long. You buy it by the linear foot, which will actually be 1 linear foot 7 feet wide to equal 7 square feet.

Although a layer of hog wire strengthens slabs, it might not do a lot to prevent cracks. Its main function is to keep cracks from separating horizontally or vertically.

Before the actual pour, be sure hog wire lays flat. No part of it should be allowed to stick up past the tops of forms. Also, notify all your helpers that they will have to occasionally reach down with a rake or their hand to pull wire off the ground and into the middle of concrete. Laid flat on the dirt, wire will not do much good. Ideally, it should rest in the middle of concrete and be completely encased.

Your final check must also include the ground base. Is it wet or is it bone dry? During hot, dry weather, base material can almost never be too wet. In extreme desertlike conditions with parched ground and temperatures hovering around 100 degrees, professionals have soaked bases to the point where concrete was poured over water puddles and they still had to work fast to finish it before it quickly dried out.

In cases such as this, you might be wise to set up and flow a lawn sprinkler over bases the night before a pour to ensure that your concrete does not "get away from you" or you "lose it"—terms used by professionals to describe slabs that dried much too fast and did not allow enough time to effect a smooth finish.

Finally, be sure to talk over the job with helpers. Let them know what to expect. If your job is such that the driver will dump half the load and then have to reposition his truck, alert helpers so they can assist the driver by guiding him in and out of tight spaces. Remember, the truck will make a lot of noise and normal voice communications will be hampered. Be especially careful when directing the driver so he does not run over forms, stakes, or helpers.

CONCRETE LOAD CHECK

After the driver has walked the job site and positioned the truck, he will get out and ask if you want to see a sample of the concrete. If he doesn't, you insist. Always check concrete before actually pouring. This gives you the opportunity to have water added to dry mixes. Dry concrete is very difficult to work with. Unless your job consists of stairs or a very steep driveway, wetter concrete is recommended.

Concrete delivery trucks are equipped with huge drums that continuously rotate to keep concrete thoroughly mixed. Welded to their inner drum surfaces are a series of spiral shelves. As long as the drum rotates against these shelves, concrete is forced toward the bottom of the drum. To get concrete to exit drums and flow down chutes, drivers simply reverse the drum's rotating direction. This causes drum shelves to force up concrete material much like a drill bit forces out curly wood slivers.

To see a small sample of concrete, the driver will reverse the drum's direction until concrete flows 2 to 3 feet down the chute. Then he will stop. Look at the mud and determine its consistency. Dry concrete will not flow down the chute easily and will tend to batch up on itself. Wet concrete will flow like water and not batch up at all. Unfortunately, if your load is wetter than you prefer, there is little you can do about it. If the mud is dry, however, you can have the driver add water to the mix. Concrete trucks carry their own water supply for this purpose.

Confer with the driver about the concrete's consistency. Make sure he understands you are a novice finisher and would prefer a wet load to a dry one. If concrete piles up on the chute and seems to have a tough time flowing, have the driver add a half gallon of water per yard of concrete. Once this is done, ask to see another sample.

Good concrete mixes flow easily down chutes. They are also much easier to screed, tamp, and finish (FIG. 9-2). When you believe the mix is just right, allow some to dump on the ground (inside forms, of course). A good mix should not pile up, rather, it should flow outward and essentially flatten itself out. In addition, good mixes will fall off the ends of chutes in a constant flow, where dry mixes tend to plop on the ground in stuttered batches.

It is not uncommon for concrete to start to dry out in the middle of a pour, especially on hot days. If this happens, stop the driver and ask that more water be added to the mix. A half gallon of water per yard of concrete is a good starting point, although a lot of finishers are confident that one gallon per yard is not excessive. Just remember, though, you can always have water added but you can never take water out.

DIRECT POUR

Pouring concrete directly from a truck's chute requires at least three, preferably four, workers. One to handle the chute and direct the driver, two to screed, and one to assist screed operations by moving concrete with a rake or shovel.

The chute operator is responsible for directing the chute in such a way that concrete fills all reachable areas with each pass. Essentially, the chute is moved from left to right on one pass and then right to left on the next, and so on. Each pass should result in just enough poured concrete to fill a swath up to the tops of forms and the bottom of the screed board. Too much will require removal by shovel or rake, a heavy work load. Too little might require another quick pass of the chute or a few shovelsful to fill in. Long, steady sweeps of the chute work best.

The chute operator needs a keen eye for estimating how much concrete falls on the ground. He must also signal the driver when it is time to

9-2 A good workable concrete mix pours freely out of the chute.

move forward or stop the concrete flow. When drivers have a good view of pours, they can tell when to move ahead and when to stop. Should the job of pouring get ahead of screed workers by more than 4 to 6 feet, stop the concrete and let them catch up. This is really important if too much concrete has been poured, as all of the workers will need to get shovels and rakes to help move excess concrete back to empty form areas. Operating a chute to result in a perfect 4-inch-thick pour takes a lot of practice. Have patience and never be afraid to stop the truck if the pour starts to progress too fast.

Although one worker will be assisting screed operations by moving excess concrete away from the screed board and filling in low spots as they occur, a chute handler can sometimes help flatten and place concrete by pushing it with his foot (FIG. 9-3). This works great for forcing concrete into corners and flattening inadvertent high spots.

If only three workers are available for your pour, the two screeders

9-3 A chute operator can help spread concrete by pushing it around with his foot. Two finishers will screed one side of the driveway after concrete is poured and roughly leveled out with rakes and/or shovels.

will have to work together with shovels and/or rakes to help initially place concrete (FIG. 9-4). Before enough concrete is on the ground to screed, they can do their best to get a smooth, 4-inch-thick slab in place. They should use a rake to push and pull concrete into a flat and even layer, the same as they would with dirt in a flower bed. When about 4 feet of mud is on the ground, stop the flow of concrete. Two workers should operate the screed board, and the chute operator assist them by pulling excess concrete away or filling in low spots with shovelsful of mud (FIG. 9-5).

Working together as a team is crucial. The screeders, chute operator, and extra helpers must communicate with each other for any job to go smoothly (FIG. 9-6). As much as possible, assist each other so that just the right amount of concrete gets on the ground to eliminate unnecessary shovel and rake work. And remember, when the job of pouring starts to get ahead of screeding, stop the concrete to catch up with screed work (FIG. 9-7). This time taken to catch up will help as much as anything else to make your job progress smoothly and efficiently.

WHEELBARROW MANEUVERS

Moving concrete from a delivery truck's location to inside forms is hard work. Concrete is heavy and it takes a little time to get used to "wheeling" it. Have the driver park his truck as close to the job site as possible to make the wheelbarrow run as short as can be expected (FIG. 9-8).

9-4   Teamwork is the key with any concrete flat work job.

9-5   Screed boards are easiest to operate with two people. Boards are moved back and forth over slabs sideways while being pulled at the same time. Note the worker in the middle removing excess concrete with a rake.

9-6 The screeder on the right is at a screed pin and cannot continue screeding without lifting the board. To ensure a flat screed, the worker on the left has continued screeding. After going about 2 to 3 feet ahead, the screeder on the left will pick up the board and move it back to the spot across from the screeder on the right. At the same time, the board is raised over the screed pin so the job can continue.

At first, have wheelbarrows filled only half full. This lighter load allows users to get a feel for moving the wheelbarrow and its shifting load. It also gives them a chance to determine the best route of travel and what it feels like to go around corners, over ramps, and through tight spots. As the job progresses and "wheelers" get used to the work, they can have the driver pour fuller loads.

If a wheelbarrow should get off course, fall over, and lose its load, all is not lost. Use a shovel to scoop mud back into the wheelbarrow or into formed areas when close by. As time permits, the concrete residue left behind by the lost load can be washed away with water. In worst cases, a wire brush works great for removing dry concrete.

9-7   This screed board was easily maneuvered on top of the concrete apron and form for the section on the left. In the middle of the driveway, it is supported on a screed form by an attached ear.

9-8   Avoid running wheelbarrows over forms. Here, a wheelbarrow has been placed inside the formed area to deliver concrete next to the house where chutes will not reach.

Dumping concrete out of wheelbarrows is done in a smooth rapid movement, much like trying to throw concrete out of it. Many novice wheelers slowly tilt wheelbarrows allowing concrete to gently plop out. Along with making a mess from plopping concrete splatter, slow-motion dumping causes concrete to pile up which, in turn, requires shovel or rake work to smooth out.

By rapidly tilting and throwing up the handles of a wheelbarrow, concrete is forced out and away to flatten and cover a wider area. In fact, done properly, the lip of a wheelbarrow will just about hit the ground first before any concrete is emptied.

Start to tip wheelbarrows about 2 to 3 feet before you reach the edge of concrete already in place on the ground. This way, concrete rapidly forced out of your wheelbarrow will reach up to, and join, that mud just in front of it instead of piling up on top of, or too far back from, the fresh concrete line.

Wheelbarrows always function best when their tire is filled with air to its proper level. Soft tires make wheeling extra difficult. This is especially important when wheeling requires going over ramps, bumps, and around corners. Ramps can be made to go over just about any reasonably sized obstacle. Plywood works good as long as 2-×-4 blocks are placed underneath for support. Going over forms requires that ramps be on both sides to keep from forcing forms out of position. Should you have to go over forms in order to place concrete, consider making ramps at two separate locations. This way, wheelers can get into formed areas from one ramp and exit the area from another to stay out of the way of oncoming wheelbarrow loads.

Because concrete is so heavy, it is best to let your arms hang down while holding up wheelbarrow handles and moving. Relying just on the strength of your bent arms for maneuvering creates a high center of gravity. By allowing your hands and arms to hang as low as possible, a great deal of the power and strength needed to maneuver wheelbarrows is shifted to your legs. With feet spread slightly apart, you'll have much better control of wheelbarrows.

SCREEDING

Screeding, also referred to as rodding off, can be accomplished in basically three different ways: with a screed board on top of forms at both ends; with the use of screed forms; or, by wet screed maneuvers. Screeding is basically just using a straight board to scrape off excess concrete from slab surfaces to flatten them as much as possible.

Walkways not much wider than 4 feet can generally be screeded by one person as long as the screed board can rest on forms at both ends (FIG. 9-9). For slabs wider than 4 feet, plan on having two people maneuver the screed and help pull back excess concrete, as well as look for low spots (FIG. 9-10).

Two people working together on a screed board that rests on forms can slide the screed board back and forth in a sawinglike motion to push

9-9 One person can generally screed areas up to 4 feet wide provided the screed board is supported on both sides. Use a hand float to move excess concrete out of the way or to scoop up small amounts for filling in low spots.

9-10 Constantly be on the lookout for low spots while screeding.

down rock and pull back excess concrete in one step. Simply pulling a screed board straight back tends to make rock tips stick up, leaving small holes behind. A helper behind the screed board and equipped with a rake or shovel, can greatly assist screed operations by alertly pulling back excess concrete or shoving needed mud into low spots (FIG. 9-11).

While screeding, be sure to keep the screed board end in contact with support forms at all times. Once in a while, rocks will land on top of forms and cause screed boards to ride on top of them instead of form tops. Operating a screed board higher than planned will cause high spots

9-11 Use a rake to pull back excess concrete. A slight angled ridge is apparent from the rake toward the far end of the screed board. This is a sign that the screeder on the left stopped at a stake while the right-side screeder continued on. After moving ahead a couple of feet, the right-side screeder stopped and picked up the board to reposition directly across from his partner while the board was raised over the stake.

along slab surfaces. This makes for an uneven screed and a surface that is more difficult to finish because of its uneven nature.

Because you must stand in concrete while screeding larger slabs (narrow walkways can be screeded from outside forms), footprints are left behind every time you move. The easiest way to remedy this problem is to reach behind with your foot and drag some mud to fill holes (FIG. 9-12). When the holes are filled, tap them lightly with the bottom of your foot to flatten them out. Then, screed over them.

9-12 Fill shallow low spots with handfuls of concrete or small scoops with a hand float. Larger areas can be filled with shovels full. Be sure to screed over all areas that receive extra concrete from any source.

Excess concrete accumulated along a screed board is quite obvious. Low spots are just as important to take care of but might be a little harder to see. Low spots are pockets of concrete that are not filled high enough to reach the bottom of screed boards. Unless they are filled and screeded flat, they will cause slab finishes to be uneven and also serve as spots for water to puddle and dirt to accumulate once the slab hardens.

Low spots are places that look like they have not been touched by a screed board and, in fact, they haven't. Normally, screeding makes concrete surfaces appear smooth and slick because cream is forced to the surface while aggregate is forced down. Conversely, low spots are rough with aggregate and plainly visible. Fill them with a shovelful of concrete or push some mud into their area with a rake, provided they are located close to the screed board. Once filled, run the screed over them to ensure flatness (FIG. 9-13).

It is not uncommon to have to screed surfaces more than once. After screeding a 3 to 4-foot section, check your work to see if any high or low spots linger. If so, go back to the area and screed again (FIG. 9-14). Most of

9-13 Pockmarks in front of the screed board are signs of low spots. After the screed board has completed this pass, workers will fill the spots with handfuls of concrete.

9-14 Use a hand float or your hand to remove excess concrete when screeding over stakes.

the time, second and third screedings are needed for pours where too much concrete was placed to begin with. As the screed board was maneuvered over the area, it was forced to ride on top of concrete because enough strength was not available to move the amount of excess concrete. This is clearly one of those situations where concrete flow should be stopped and all hands used to move excess concrete away from the screed area. Concrete should not flow again until the entire area is screeded properly.

When stakes stick up above form tops, screed boards cannot smoothly pass by. The best way to screed around them is to have the opposite end of the board screed go past the obstruction at an angle. Once the other side has passed the stake a couple of feet, lift the screed board over the stake and have the person handling the opposite side move just ahead of the point where you stopped (FIG. 9-15).

9-15 When obstacles prevent normal screeding, use a hand float to flatten and level out concrete. Here, concrete is carefully placed around a hot tub railing.

By having the opposite end of the screed board go past the stake, you are guaranteed that the area was actually screeded except for a very narrow triangle out around the stake. This small area can then be easily flattened and smoothed through tamping and bull floating.

Should large obstructions prevent normal screeding, try using a smaller screed board. In very small spaces, you can use a hand float to screed. This will require a keen eye for detail and level. Each circumstance is different. Use your ingenuity to solve special screed problems. Don't forget that chalk lines can be useful, especially when placed above the level of concrete so they are not covered during the pour.

Operating a screed board along screed forms presents some extra precautions. Be aware of the screed board ear at all times. Should it slip off of the screed form, it will cause a long divot in the concrete surface. Mud forced out from the divot will be higher than it is supposed to be and will require the area to be completely re-screeded. Another precaution involves screed forms that are suspended over slabs and only supported every 4 feet or so by stakes or screed pins. Because the screed form is suspended, excessive downward pressure on the screed board could cause the form to bow. This allows the bottom of the screed board to fall below expected slab surfaces and causes low spots. To avoid this, concentrate on pulling screed boards back only and not applying any unnecessary downward pressure.

Suspended screed forms should not be buried in concrete. In fact, the only part of them that remains close to the surface is their bottom edge. Therefore, only stakes or screed pins should have to be pulled from concrete and only their holes left behind should need to be filled. Because these holes will be rather small, you can wait to pull them until operating a tamp close by. There is no compelling reason to walk out on fresh concrete just to retrieve screed forms and stakes when you will have to walk out there to tamp soon anyway. Besides, you can easily cover and flatten your footprints with a tamp.

A tamp operator will need help getting screed forms and stakes out of the slab area. He can loosen stakes with a hammer and then pull forms with stakes attached all in one action. Helpers will have to stand by next to forms to get the board from him. If the boards are too long, he might have to remove nails, hand the board to you or a helper, and then retrieve stakes separately as he comes to them. If need be, partial shovelful of concrete can be handed to the tamp operator to fill holes as needed.

Wet screeding

Wet screeding is done when forms are not available on one side of a slab to support a screed board. In addition, screed forms might not be available because of space limitations, the amount of work involved to place them, or any number of other reasons. Because they have plenty of experience, many professional concrete finishers prefer to wet screed along house and other structure walls when pouring 3 to 4-foot-wide walkways. The reason is simple— they can achieve an excellent screed without having to perform the extra work involved with screed-form construction.

To wet screed, first snap a chalk line against the wall you will pour next to. This will designate the concrete's top surface. In addition, snap a line $3^{1}/2$ inches above the first line to use as a reference for the top of your 2-×-4 screed board. Be sure that the lines are straight and conform with the grade you have established.

While pouring concrete, use a hand float to position concrete along the wall, extending out about 8 inches to a foot. Use a hand float to create a smooth flat strip of concrete with its edge on the lower chalk line (FIG. 9-16). It is this smooth and even strip of concrete that will serve as a screed board guide along with the upper chalk line. Work on this strip until you have it correctly placed for about 6 to 8 feet along the wall.

You will have to direct all of your attention on the chalk line and flat concrete strip at the same time while working with a wet screed. Because you cannot easily look for high and low spots, have the screed board worker on the other end of the board and the helper take care of them.

Use a firm hand to glide the screed board over concrete while following the upper chalk line. With two guides to work from, there should be no significant problems (FIG. 9-17). The sawing-type motion suggested for

9-16 Concrete is floated up to the designated chalk line for a distance of 8 to 10 feet before actually screeding. A lip created by the plywood wall siding helps this finisher to accurately float concrete into place.

9-17 Operating the free end of a wet screed requires total concentration.

regular screeding cannot be used when wet screeding. That much board movement makes it almost impossible for a wet screeder to maintain a flat screed posture.

As with other screed work, you have to go back over areas that have divots, low spots, and other imperfections. Do the job right, no matter how many attempts it takes. Although wet screeding is easiest on narrow walkways and small slabs, professional finishers have had good results wet screeding up to 8-foot widths. Their success in this endeavor is based on experience, patience, and excellent screed worker communication. Novices are advised to rely on screed forms for any job wider then 3 feet.

OVERVIEW

Whether your job is screeded on top of form boards, screed forms, or just wet screeded, the end result is what's important. Pouring concrete and screeding it off requires a lot of physical labor. By the same token, these concrete placement operations will determine, to a great extent, how well your concrete pour turns out. No amount of finish work can remove excessive high or low spots. Once you call the screed job good, that's it. All the rest of the work will involve smoothing the surface and making it look nice. Therefore, have plenty of help available for the pour, fully explain the job at hand, do not be afraid to stop the flow of concrete to catch up screed operations, and do not get in an uncontrolled hurry. If the mud starts to dry out, add water. If a form breaks loose, stop and repair it. By all means, take care of problems immediately or direct someone else to do repairs for you. Have all of your equipment ready at hand, including screed boards that fit across forms adequately. If the job includes a narrow walkway and a larger patio slab, have a separate screed board available for both so you won't have to hassle with a long screed board just to screed a narrow walkway.

Chapter 10

The first hour

*T*he first hour after concrete has been poured and screeded will be busy. Along with tamping and floating, you might have to insert stirrups or J-bolts, anchoring devices used to support patio cover posts or walls. In addition, you'll have expansion joints, edges, and seams to work on; screed forms to pull; and overflow concrete to remove from adjacent, preexisting slabs or yard areas. Some cleanup operations, such as rinsing wheelbarrows, shovels, and screed boards, can also be started. Although thin layers of concrete residue will harden faster than slabs, you can delay tool and equipment cleaning until after tamping, bull floating, and other initial concrete finishing chores are complete. However, if plenty of helpers are available, have one start cleanup operations early on to be sure that concrete residue is completely washed away before it sets up.

TAMPING

Optimally, tamp work should start as soon as concrete has been screeded. With enough helpers on site, one could begin tamping right behind screeders. For all intents and purposes, tamp work can never be started too soon after screeding. In fact, the earlier you tamp, the easier it will be. This is because concrete is still wet enough to allow aggregate to be easily pushed into the mix. Waiting too long lets water and moisture evaporate from slabs to make the mix hard and, in turn, make it more difficult to force aggregate down from the surface.

Tamps are designed strictly for use on concrete surfaces to push down aggregate and bring up creamy concrete mud (FIG. 10-1). Wire mesh screens on the bottoms of most tamps usually measure 3 feet long by about 8 inches wide. Because of this relatively small size, tamp users have to walk around wet slabs in order to reach every square inch of the surface. Working in a backward direction, tampers fill in footprints by tapping around them with the bottom of their foot. Once initially filled in,

10-1 Drop tamps on fresh concrete to push down aggregate and bring up cream. This tamp has been operated correctly as indicated by the uniform pattern left behind.

they simply tamp over the spots to flatten and smooth them over (FIG. 10-2).

Fresh concrete is easy to tamp, as there is plenty of moisture in the mix to allow aggregate to be forced down and covered with a solid layer of cream. Conversely, concrete that has sat around for a while gets stiff and dry, and it will have to be forcibly pounded with a tamp over and over to achieve a smooth and rock-free surface—a very labor-intensive operation.

The same dilemma can happen if concrete is poured over a dry dirt base, especially on a hot day. All of the water in fresh concrete will be rapidly absorbed by dry dirt and quickly evaporated from the surface, leaving behind stiff concrete that is almost impossible to tamp with any success. To avoid that situation, be sure base material is thoroughly soaked with water and that tamping begins just as soon as screeding maneuvers allow.

Unless concrete is very dry, a great deal of downward tamping pressure is not necessary. The weight of a tool alone dropped from just a few inches above the surface is generally enough to provide a sufficient job. Done correctly, tamping results in a flat slab with no holes or protruding rocks but with a wafflelike surface finish. This pattern is created by the wire mesh screen on the bottom of the tamp. This wet, bumpy texture is erased and smoothed with a bull float (FIG. 10-3).

While operating a tamp, maintain the screen mesh in a flat plane,

10-2 Footprints are unavoidable while tamping. Lightly tap footprints with your foot to fill in areas and then tamp over them to flatten surrounding concrete.

10-3 Wide bull floats do an excellent job of smoothing bumpy-textured concrete surfaces after tamping. Bull float as soon as possible to ensure concrete is wet enough to accept easy smoothing.

even with the surface. These tools are designed to make the job comfortable for the operator at any angle and easiest to apply in flat motions. If they are tamped against concrete surfaces with the front or back edge hitting first, a wavy and divot-marked surface texture will result. This type of inconsistent and uneven surface texture is difficult to flatten later with a bull float and fresno applications.

You can determine the flatness of your tamping efforts by evaluating the tamp's screen mesh pattern. If it is not evenly dispersed, adjust the position of your hands on the handles either toward the front or back until you start tamping evenly.

Screed work frequently leaves behind very slight ridges running across slabs in the direction boards were operated. To help flatten ridges and perfect surface flatness, tamp in the opposite direction of screeding. If you moved the screed board from east to west, then tamp from north to south. Tamping should start where the first concrete was poured, operated over 2 to 3-foot areas and then backward. Fill in footprints by tapping around them with your foot, then flatten spots with the tamp.

Once you have made a pass from one side of the slab to the other, simply turn around and work back the other way. You will tamp back and forth until the entire slab surface has been covered. Reach out as far as possible while keeping the bottom of the tamp flat to the concrete surface. Allowing one edge of a tamp to contact concrete harder or sooner than the other will cause minor divots on the surface, which are difficult to smooth later on. Tamps are designed with a balance factor to help users apply them flat and evenly. Let the weight of the tool do most of the work. Trying to force it into concrete makes the job harder to accomplish and usually results in uneven surfaces.

Many experienced concrete finishers have noticed best results along form edges when they apply tamps with their 3-foot front side in line with forms. This longer tamp edge is easier to guide next to forms and also does a better job of tamping. Do this along all form edges, including stringers.

If your concrete slab has suddenly gotten very hard and dry before tamping—because the base was not wet down sufficiently or it is extremely hot outside—you might really have to pound down on the tamp with every blow. This is done to knock rock and aggregate down, bring up all available moisture and cream, fill in tiny crack voids surrounding surface rock, and to create enough cream for finishing tools to work with. Should extra hard tamp pounding not work well, you might be forced into sprinkling a fine mist of water over the slab surface.

Although adding water to concrete finishes tends to dilute mixtures and might weaken surfaces, a very fine spray lightly applied to hardened surfaces is about the only way to make them soft enough to accept a decent tamp job. In these situations, you must have someone operate a bull float over the slab immediately after each tamp pass. If not, you will have an exceptionally tough time smoothing the wafflelike pattern left behind by the tamp. By all means, avoid this dilemma by soaking bases with plenty of water before pouring and by having enough helpers available so tamping can take place immediately after, or during, screed operations.

BULL FLOATING

Bull float tools are 3-foot pieces of wood or metal about 1-inch thick and generally measuring around 8 to 10 inches wide and are equipped with adapters for pole extensions because you cannot walk on slabs at this point. They are used to smooth wet concrete surfaces immediately after tamp work is completed and from various angles to fill in cat's eyes— small and shallow low spots that a bull float never touched. The weight and wide surface area of bull floats work together to smooth tamping bumps and bring up moisture.

Most professionals initially operate bull floats from a position perpendicular to that of tamping. In other words, if a screed board was pulled over a slab from east to west, tamping would run in a north/south direction to help flatten out any minor divots or dents produced by jagged screeding. A bull float is then operated in an east/west manner to compensate for any tamping flaws, filling in shallow low spots in the process. If need be, of course, you can maneuver a bull float in any direction to pick up cream and fill in low spots.

Bull float work is done immediately after tamping. In fact, if enough helpers are available, one should bull float while another tamps and others complete different tasks. The sooner tamp and bull float work is accomplished, the better. Both of these operations bring up lots of cream to the surface, and it is this cream that will ultimately be finished to give your concrete slab surface its final appearance.

Waiting too long to float makes smoothing jobs difficult. Instead of a lot of wet cream to work with, moisture will have evaporated leaving behind tiny cracks and pockmarks that cream has sunk into (FIG. 10-4). This condition requires you to bull float over the surface many times and from all directions, essentially working up cream from wet areas and floating it over to dry ones. All holes will have to be filled. The entire surface must have a somewhat smooth texture before fresno work can be done.

Because concrete should still be wet, lines and ridges will be left behind from bull float edges. If these lines and ridges are so prominent that they protrude more than a half inch, you should bull float a second time, about 5 to 10 minutes after completing the first float. Don't worry if the surface finish does not remain perfectly smooth. It is normal for some smooth, slightly bumpy looking patterns to appear as moisture evaporates. Fresno work will take care of that.

Dry concrete conditions require that fresno work begin immediately, just as soon as you are done with the bull float. There is no time to waste between operations. Likewise, you will probably have to hand trowel finish right after the fresno. In other words, fast-drying concrete requires immediate and constant work. As soon as one procedure is accomplished,

10-4 Concrete in this area has dried somewhat, making bull float work more labor-intensive. Operate bull floats back and forth across dry surfaces to work up cream, fill in holes, and smooth tamping marks.

the next one must start until all procedures have been completed and the slab is completely finished. Under normal conditions, though, you should have time to rest between bull float applications and fresno work and again after the fresno and before hand trowel maneuvers.

Because extension poles might have to reach up to 18 feet, sometimes more, you might have to make room for them around slabs or figure out alternative methods for applying the bull float and fresno. This is especially significant for slabs that are bordered on two or three sides by walls, fences, structures, etc. (FIG. 10-5). You might have to float and fresno half of a slab from one side and the other half from its opposite side.

Even if you have room to operate with more than three, 6-foot extension poles, you will have trouble. Poles extending over 18 feet tend to bow in the middle, making it almost impossible to raise them high enough so the back edge of your bull float or fresno can be raised off of the slab surface when the tools are pulled back toward you. Be sure to consider factors like this before the pour. Make arrangements to remove fence sections or make other adjustments so that all parts of your slab can be easily accessed by the bull float and fresno.

Adjust the bull float adapter swivel so that the tool is pushed away from you with its front edge off of the surface to prevent it from digging into the concrete. This will entail holding poles low to the ground (FIG. 10-6). The swivel adjustment should also be positioned so that you are

10-5 Slab areas bordered by obstructions might require bull float work from a number of different directions.

10-6 Heavy lines left on the surface as a result of bull float maneuvers are a sign that concrete is still wet.

able to raise poles high enough to lift the back edge of the float while pulling it back toward you so it doesn't dig into the concrete. Simply hold poles low while extending the float out onto the slab and then raise them high to bring the float back.

Ideally, bull floats should be operated over slabs in the opposite direction of tamping, just as tamping was done opposite of screeding. Working in opposing directions helps to better smooth and flatten concrete surfaces, as well as fill small low spots and other imperfections. Obstructions can prevent you from operating a bull float in the opposite direction of tamp work. This is common and you can certainly end up with a fine looking slab even though the tamp and bull float were operated in the same direction. It is just that working in opposite directions can provide a little better coverage, like mowing lawns in one direction and then the other to guarantee every blade of grass is cut.

Staying ahead of the concrete is the name of the game. At no time will you want concrete to dry out and get hard before you are ready for final hand trowel finishing. Pay close attention to timing and start each successive work phase right away. This first hour of working concrete will greatly affect how smooth the rest of the work goes.

HAND FLOATING

Hand floats are small trowels made of wood. Like bull floats, they are used to initially place and smooth rough concrete. Steel hand trowels are used to actually effect final finishes. Hand floats are useful in placing mud into tight spaces while pouring and screeding, flattening out rough spots, and smoothing surface material. The front portion of the tool works great as a scoop to place small amounts of concrete under mud sills, lips formed by wall siding, and lots of other limited access spaces. While screeding, many finishers keep a hand float nearby to assist in moving excess concrete away from screed boards.

Form and stringer tops are quickly cleaned by scraping a hand float edge against them (FIG. 10-7). You can also use this tool to remove concrete residue from posts and other obstacles inside a concrete area. Hand floats work much better than your hand to smooth and reposition concrete simply because they are flat.

Hand floats are often used to smooth rough, wet concrete along slab edges close to forms (FIG. 10-8). Bull floats are big and not always easy to maneuver along edges. Slab perimeters generally dry quicker than the middle because moisture is lost at the sides as well as the tops and bottoms. Therefore, hand floats might need to be used along slab edges to scrub concrete and bring up creamy finishes. Hand float wooden blade textures work great to force down aggregate and bring up workable cream slurries. Once a dry area has been scrubbed to a creamy finish, a couple of light hand float swipes should smooth it in preparation for a fresno (FIG. 10-9).

Places too difficult to reach with a bull float might be accessed with a hand float. You will not be able to walk on concrete after it has been

10-7 While other work is being tended to, extra helpers can clean form tops with a hand float for a clean working guide while edging.

10-8 Scrub and smooth rough concrete along the edge of forms with a hand float.

10-9 Once rough concrete has been sufficiently scrubbed to bring up cream and fill holes, a few light passes should smooth it with an even texture.

tamped, but you might be able to reach out-of-the-way spaces by walking along walls, forms, steps, or fence rails. Hand floats are miniature bull floats. Their design and construction is intended to offer the same results as bull floats but with much greater maneuverability (FIG. 10-10).

The perimeters around large obstructions in the middle of slabs are easily smoothed with a hand float. Items such as manhole covers, water meter lids, and so on, are difficult to work around with large bull floats. Hand floats are used for a lot of various tasks. While checking and smoothing slab edges, clean off excess concrete from the tops of forms so that mud does not accidentally get knocked onto slabs during other work.

SEAMING

Seams, also called control joints, are made on concrete surfaces in hopes that any eventual crack will occur inside the seam rather than diagonally across nice finishes. Although seams are most common on walkways and sidewalks, large driveways and other big slabs should have them too. Used in conjunction with expansion joints, seams provide a valuable means by which to control cracks.

Walkways should have seams about every 3 to 4 feet with an expansion joint every 10 to 12 feet but it depends on the walkway length. Seams should be placed so that they are evenly spaced and symmetrically

10-10 Because of the round porch, a regular bull float was not able to fill in concrete around bricks or smooth in tight spaces. A hand float is used to accomplish what the large bull float could not.

patterned. Seams and expansion joints are needed for walkways because of their generally long and narrow shape. Their length does not allow room for concrete expansion on hot days and the narrow width doesn't provide enough concrete to add necessary strength or support.

Driveways should be equipped with a seam every 10 feet and an expansion joint every 20 feet running across their width. Unlike patio slabs that usually do not have to support a great deal of weight, driveways are consistently subjected to heavy vehicles. Heavy weight loads increase the chance of cracks and seams help to control this.

Seams are placed using a 2-×-4 board as a guide. The board should be long enough to span the slab, preferably placed so that it is supported on both ends by forms. Nails strategically placed along outer form faces are great for designating the location of seams. This should be done before the pour when you have plenty of time to measure the job and determine the most accurate spacing.

Seamers have a definitive ridge in the center of the blade. This ridge is what causes a seam groove on concrete surfaces. Because this ridge is located a few inches away from the tool's outer edge, you must position 2-×-4 guides appropriately. In other words, if you put a seam at the nail guides, you must position your 2-×-4 guide away from the nail far enough to account for the distance between the seamer ridge and the outer edge. If the 2-×-4 edge is placed in line with the nail, the seamer puts the seam a few inches away from the nail's position (FIG. 10-11).

10-11 A long, 2-×-4 board is used as a guide for an initial seam installation. Because this driveway is so wide, a walking seamer is needed.

Seams are easiest to apply while concrete is wet. The soft texture of wet mud allows seamers to smoothly glide across surfaces to quickly create deep, crisp seams. Waiting too long to effect seams makes pushing aggregate down and forming cream into a crisp groove difficult. Therefore, plan to initially install seams right after concluding bull float maneuvers.

After making seams, remove the 2-×-4 guide boards. You might need help on long ones; they sometimes stick to wet slabs with a suctionlike grip. Use a bull float to remove lines or imperfections left by the board. After an initial seam has been made, don't worry about it filling with cream. Your main purpose was to move rock and grit out of the groove. Fresno activity will fill seams with cream, which is easily reshaped with subsequent seamer applications (FIG. 10-12). Plan to run seamers along seams after every other fresno application.

You should not have to use a 2-×-4 guide every time you use a seamer. Guides are only necessary while initially making grooves (FIG. 10-13). After the first application, an indentation has already been established and cream will not cause seamers to wander like rocks would. Take your time and maintain a firm steady motion. Lines to the sides of seams left behind by seamer edges are wiped out with a fresno or hand trowel. As the slab hardens, use a seamer to simply finish groove surfaces as opposed to actually shaping them. A light touch is best. Once you're done, be sure to wipe out edge lines with a trowel.

Walkway seams are easily made using a hand-held seamer. For longer

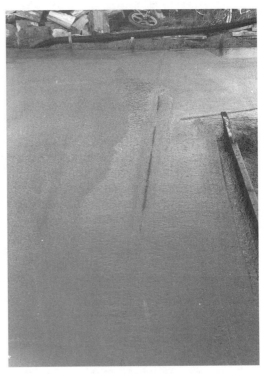

10-12 Fresno applications done after initial seams have been made will fill in grooves with cream. A second seam application will not be hindered by aggregate and cream will conform to the desired shape.

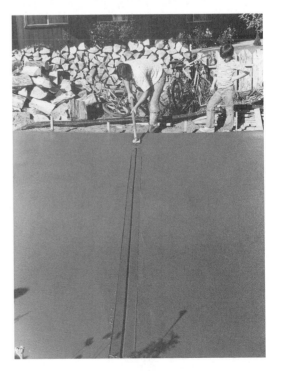

10-13 Because aggregate was pushed down early, this second seam application can be carefully completed without a 2-×-4 guide.

10-14 A corner has been formed by pouring around a preexisting concrete apron in front of a garage. This seam should control cracks in this area. It is important to install seams and/or expansion joints anywhere corners are formed.

seams on driveways or sports courts, you'll need a walking seamer. These larger tools are available at rental yards. Their adapters should conform to those on your bull float and fresno so that the same extension poles can be used on all three tools.

In addition to placing seams at strategic points across slabs and walkways, you should insert them at every corner (FIG. 10-14). Each time a walkway makes a turn, the corner of it should have seams inserted out from its inside corner. If seams or expansion joints are not provided, that corner of the walkway will crack (FIG. 10-15).

EDGING

Edging wet concrete soon after bull floating has the same effect as putting seams in early; it gets rocks out of the way while mud is still wet and soft. And, like seams, these edges will probably get filled with cream from fresno operations. Plan to go over edges after every other fresno application.

Walking edgers make quick work of edging concrete (FIG. 10-16). The tool's rounded side is inserted between a form edge and concrete. Its flat part will smooth concrete away from the slab edge while the rounded section actually effects a curved corner along concrete edges. As with other concrete tools, you have to keep the front part up and out of concrete while moving forward and the back part up when going backward (FIG. 10-17).

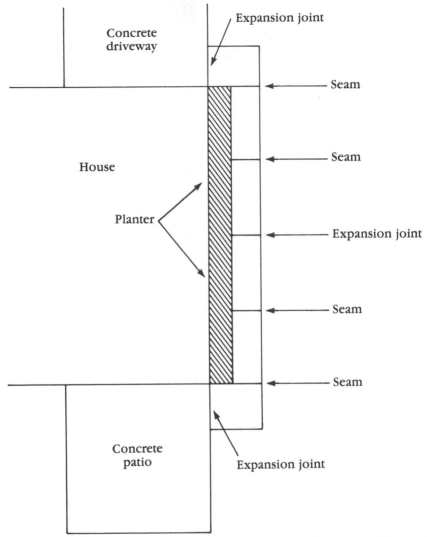

10-15 A walkway poured against a patio slab in the back and a concrete driveway in front should be equipped with seams and expansion joints as illustrated.

Forms serve as guides while operating edgers. Gently allow the side of your edger to run against forms while keeping a close eye that the tool does not stray off course. Concrete's newly formed round edges should be smooth, but go over them again if holes or blotches appear. Even though this might be just an initial edging, try to shape and smooth edges correctly. This way, later maneuvers will be easier to effect and finish.

Hand edgers must be available for most concrete pours even though a walking edger is on site. These are used in tight spots where walking models might not fit. They are also great for repairing edges that might

10-16 While moving forward, be sure an edger's front edge is raised off the surface.

10-17 Operated smoothly and with balanced pressure, hand edgers do a fine job of making crisp rounded corners. Remember to keep the leading edge raised off the surface to avoid making divots on the surface.

eventually suffer nicks, dents, or other blemishes. Hand edgers are much more maneuverable than walking ones and better suited for intricate work.

If you do plan to use both types of edgers, make sure they both have identical, rounded-arc dimensions. Edgers are generally available in 1/4-inch, 3/8-inch, and 1/2-inch arcs, with the 3/8-inch model seemingly the most popular.

Hand edgers are essentially operated the same as their larger counterparts. You must raise the front end while moving forward and the back end when going backward. Apply even pressure to edgers. Too much force on a rounded side tends to push an arc's edge into a slab's side, making an awkward concrete edge. Too much pressure on the edger's flat sides cause deep lines on concrete just off the tool's outside edge. Evenly applied pressure effects crisp, round concrete edges with hardly any line from the flat side.

Lines left on concrete surfaces are wiped out with hand trowels. However, extra deep lines continually made with each edger pass require significant amounts of trowel work. This is not good, because grit will eventually be worked up onto the surface as edges become more uneven (FIG. 10-18).

10-18 The purpose of an early initial edging is just to make sure aggregate is pushed down and out of the way. Overworking edges results in heavy lines that must be hand floated while concrete is still wet.

All concrete surfaces that come in contact with forms, expansion joints, and preexisting concrete flat work should be edged. This includes both sides of stringer tops as well. One of the few times when edges are not needed are along house and other structural walls. Concrete is finished flat against them.

FIRST FRESNO

Your first fresno application should be made when a concrete surface has lost most of its watery sheen (FIG. 10-19). Metal-bladed fresno tools are basically 3-foot-long finishing trowels that bring moisture up to slab surfaces while further smoothing an existing finish with each pass. For the most part, first fresno applications are done about 10 to 20 minutes after bull float work is complete.

10-19 Darker, shinier concrete is a sign that moisture is evaporating from the surface but that fresno work is bringing moisture up. Lines left by the round fresno blade indicates that this concrete is still wet.

Although fresno passes made too early might bring up excessive moisture, leave ridges behind from tool edges, and possibly bring up some sand particles, it is better to fresno too early than too late. If you are in doubt as to how long to wait, try the fresno on a slab section to test the results.

A good fresno timing indicator is the watery sheen on slab surfaces. Once the sheen begins to dissipate, it is usually time to fresno. On very hot days, you should apply a fresno right after bull floating. Hot weather causes concrete to flash, which means it dries out quickly and suddenly. Cool, damp weather tends to keep slabs wetter for much longer periods (FIG. 10-20). Sometimes, cooler conditions allow finishers to wait up to an hour before using a fresno.

When applied at just the right time, fresno applications smooth slab surfaces considerably. As moisture levels in concrete start to drop, how-

10-20 Fresno applications are made as watery slab surface sheen's disappear. As concrete gets harder, fresnos do a better job of smoothing concrete.

ever, surface textures once again take on a rougher appearance. Each succeeding fresno application further smooths the surface. In fact, it is not uncommon to fresno slabs four and five times.

Operating a fresno with extension poles can be awkward, so take care not to trip over obstacles or knock things over with the poles. Most finishers simply stand next to forms and use a hand-over-hand method to push a fresno across a slab. To pull it back, they raise poles and walk backward. You could pull it back with just your arms but that jerky motion will result in surface imperfections. Walking backward works best.

A fresno is operated just like a bull float. You must hold poles low to the ground while pushing it away and up high when pulling it back toward you. The front of the tool must be raised while moving forward and the back edge must be raised while being pulled back.

Fresno adapters are also adjustable. Make any necessary adjustments to the swivel before you start. On long slabs, be absolutely sure the blade is adjusted so that you really have to hold poles low to get the fresno out and away. This way, you should have no problem raising it high enough to bring it back without its back edge scraping on the concrete. You can get yourself in a heck of a bind if the blade is adjusted too high and you can't raise the pole extensions high enough to get the fresno off the slab without using a stepladder.

Fresno adapter swivels generally allow for sideways adjustments as well as for up and down directions. This allows you to use a fresno at an angle in tight spots. An example would be a slab poured next to a fence. If the fence blocks the extension poles coming straight off the slab, maybe there would be room if they came off at an angle. Using a fresno at an angle works great while walking along and finishing a 3-foot, 1-inch walkway lengthwise.

Chapter 11 describes successive fresno applications thoroughly. For now, though, remember to apply your fresno in a smooth and steady motion with the front edge up while moving forward and the back edge raised while pulling back. As long as your slab does not flash and the weather is a mild 70 degrees, you should be able to stop and catch your breath after the first fresno—providing all of the other, previously described chores have been completed.

MISCELLANEOUS TASKS

Removing screed pins or stakes can be difficult. Although designed to be pulled out easily, you can pound them into the ground so hard they resist easy removal. The best way to loosen them, and also the best way to prevent wrenching your back, is to smack them with a hammer. Hit them on one side and then the other to wriggle them loose. On soft soil bases, some finishers easily loosen pins and stakes with a shovel (FIG. 10-21).

Cleaning forms and tools is a task that should not go unattended. Water applied at a high pressure usually works great. Just be sure water does not get sprayed on new concrete. Use a stiff brush or foxtail broom to help loosen and remove stubborn patches of concrete residue.

10-21 In soft ground, you might be able to reach out with a shovel to loosen screed pins or stakes. Most of the time, though, you'll need to use a hammer.

Screed boards, screed forms, screed pins, stakes, and tools should be cleaned as soon as possible. Wet concrete is easier to remove than dried on remnants. Tools that will be used again must also be cleaned between applications. Hand trowels, seamers, and floats can be placed in a five-gallon bucket partially filled with water. Use a brush to clean them. Be very careful of the sharp edges on steel trowels and edgers. They can become razor sharp in no time after being ''honed'' on concrete.

During the first hour, stirrups and J-bolts should be installed. You have already learned how to designate their prospective locations (nails on the sides of forms). When placing more than one, use a string as a guide to make sure they are inserted in line with each other. Install them right after the bull float while concrete is still soft and wet. You might have to use a hammer to help persuade them into position. Recheck their position after installation to be sure they are straight up and down and in line with each other.

During the first hour, 2-inch-wide benderboard expansion joint strips must also be installed, providing you planned for them instead of felt or stringers. Place them after the first bull float. Use a hand float to smooth concrete around them afterward.

By the time you get all of your projects completed, it will probably be time to apply a second fresno. The first hour of any concrete job is always a busy one. Maybe you should have a list posted at your job site describing all of the tasks you plan to accomplish? This check list should ensure you don't forget anything and might serve you well as an overall guide.

Chapter **11**

Basic
flat work
finishing

*B*asic flat work finishing entails work with a
fresno, edger, seamer, and metal finishing trowels. At this point, providing
previously described preliminary work has been completed correctly,
your primary focus will be strictly confined to slab surface smoothing.
Concrete should have already been properly placed, screeded, and
tamped. In addition, first edges, seams, and a first fresno should have
been applied. Also, extra installations such as expansion joints, stirrups,
and J-bolts should have been taken care of.

Remember that all outdoor concrete flat work must provide some
sort of traction when it gets wet from lawn sprinklers, rain, washing, and
so on. The easiest and most common method of providing concrete surface traction is by applying a broom finish. Along with providing adequate traction for wet concrete, brooming tends to cover and hide many
small finishing imperfections.

Be prepared to apply this finish by having a soft-bristled push broom
handy. If your slab is so large that you will not be able to reach all parts of
it with your broom's regular handle, have it set up to accept bull float and
fresno extension poles.

FRESNO FINISHING

Since you should have applied a first fresno during the first hour after
pouring concrete, you should have a feel for how the tool is operated.
Successive applications are no different except that you might have to use
increased downward pressure to smooth concrete that has gotten harder.
Most of the time, additional pressure is not needed until the third or

fourth fresno. However, very hot weather can cause surfaces to flash, requiring extra power for smoothing.

Generally, second fresno applications are made when the watery sheen has once again left a slab's surface. The very top layer of concrete could still be wet and creamy, but its visible surface will have experienced some moisture evaporation and appear dull. Ripples on the surface of slabs, which are smoothed out with fresno applications, are an indication of moisture evaporation.

Run your fresno across the entire slab. As you do this, you'll notice that fresno-covered finishes become darker than untouched parts. This is normal, and is caused by the fresno's steel blade bringing up moisture from just under the surface (FIG. 11-1). As long as your fresno glides easily over slabs and brings up moisture, you are assured that the surface is still wet. Although it is not always good to overwork slabs, you should attempt to apply a fresno every time surfaces loose their sheen. If you start to fresno a dull surface only to find that the slab is still "soupy" (very wet and creamy), just stop with that pass and allow concrete to set up for another 10 or 15 minutes.

If, however, the slab loses its sheen and you neglect to fresno, especially on hot days, the delay could cause fresno applications to create pockmarks, slight holes made when bits of dry concrete stick to the blade or by excessive cream moisture evaporation (FIG. 11-2). To fill in these holes, you must operate a fresno across the slab a number of times. Bring cream and moisture to those spots by maneuvering the fresno at angles to holes so cream can be brought in from surrounding areas. Work the

11-1 The light tinted concrete in front of the blade shows that this spot has yet to be touched by the fresno blade. Notice the darker surfaces that have already been smoothed by the fresno—indicating that cream was brought up.

fresno back and forth over bad spots using whatever downward pressure is necessary. This added pressure on the fresno helps to scrub slabs and bring up moisture. Once an area is scrubbed, run the tool over it lightly to effect a smooth, even finish (FIG. 11-3).

As with previous bull float and fresno applications, you might have had to remove an extension pole or two in order to reach tight spots. Work the tool back and forth to create a smooth finish. This is really

11-2 Run the fresno back and forth over rough surface spots to bring up moisture and cream to fill small holes and pockets. If necessary, operate tools from a different direction.

11-3 A fresno wipes out scrubbing maneuvers with one smooth gliding pass. Make the last pass a full one; side to side in one nonstop motion.

important for low spots (cat's eyes). These areas will obviously appear as spots that the fresno did not touch. In fact, they were not touched, as evidenced by the darker concrete color surrounding them. Work the fresno around these areas like you would for holes or pockmarks. Sometimes, slight twisting pressure applied to poles will cause one side of a fresno to slide into shallow low spots to finish them. Use this maneuver with scrubbing action to bring up cream and fill low spots (FIG. 11-4).

11-4 A low spot is located just behind the left side of the fresno blade. As the tool passes over the spot again, the operator will twist the handle to force extra downward pressure on the left side.

The more concrete sets up and hardens, the less a fresno will affect smoothing and finishing. When concrete is just about ready for hand trowel finishing, a fresno will make a distinctive ringing sound—in effect, metal rubbing against a hard surface. Fresno as best you can with downward pressure. A lot of finishers add weights to the tops of fresno blades to help them finish harder slab surfaces. These weights actually work in unison with downward pressure applied to extension poles. The added force of weights allow finishers to fresno slabs one last time before having to get out on slabs to hand finish.

Not all slabs require the use of fresno weights. More of a luxury than a necessity, weights simply let finishers get slabs a bit smoother before having to hand finish. The smoother a slab is before hand trowel operations, the easier it is to hand finish (FIG. 11-5).

Most concrete slabs require at least three fresno applications. Timing is important. Although you should wait until watery sheens have gone, do not wait longer than 20 minutes in summer and 30 minutes in winter to fresno again. It is much better to apply an extra fresno than to wait too

11-5 Weights on this fresno are working to bring up moisture and smooth the surface over an installed seam. This will be the last fresno application, as concrete is just about hard enough to support knee boards and hand trowel finishing.

long between applications. Quality fresno work will go a long way toward reducing the amount of hand trowel work required for professional finishing results. The work accomplished with a fresno is work that will not have to be done by hand.

Deciding when a fresno is no longer useful can be tricky. The best you can do is watch and listen. Look to see how much actual smoothing a fresno is doing and listen for the distinctive metallic ringing sound made by blades rubbing against hard surfaces. This ringing noise signals that a slab is about ready for hand trowel finishing. To guarantee that concrete is ready for hand trowel finishing, test areas along edges next to forms. If you are able to effect a really smooth and blemish-free finish with metal hand trowels that also make a metallic ringing sound, the slab is ready (FIG. 11-6).

A final word about fresno work. Too many novices are afraid of ruining slabs with fresno applications. It is true that excessive fresno work will bring up sand and grit and too few fresno passes will make hand trowel work really labor-intensive. But, if you use common sense and realize that the tool is simply designed to reduce hand trowel work, you should be able to use it to its full advantage.

Obviously, if you push and pull a fresno across a slab and do nothing but move soupy cream from one side of the job to the other, you are wasting time and valuable cream. Take a break, clean some tools and let concrete set up. On the other hand, if a fresno does a good job of smoothing

11-6 Begin hand trowel finishing along slab perimeters, which can easily be reached from outside forms.

the surface, brings up moisture as determined by surface color, fills in small holes and smooths imperfections, it is working as intended.

As long as the tool smooths concrete and improves texture, great, keep using it. If only one spot is giving you trouble, work on it. Once a fresno fails to make much of an impression, it is probably time to grab your knee boards and hand trowels.

HAND TROWEL FINISHING

Most concrete slabs start setting up along their outer edges first. Therefore, test finish readiness along perimeters first (FIG. 11-7). Use one trowel to lean on and the other to reach out and work the surface. By reaching out as far as possible, you eliminate having to use knee boards in those areas. Proceed around the entire slab, leaving just a space for entering and exiting the middle (FIG. 11-8). Although most hand finishing starts where concrete was first poured, and likely the place quickest to harden, other finishers might spread out along edges to finish other sections.

Operating hand trowels does not require a great deal of experience. Although professionals can do this work much faster and with fewer mistakes, novices can accomplish the same goal by taking their time and concentrating on the overall objective—a smooth, line-free, hole-free surface. Pool trowels (those with rounded front and back sides) are highly recommended for novice finishers. These tools do a great job of finishing and

11-7 The first concrete poured is usually the first to harden; it has had more time for moisture to evaporate.

11-8 Long, wide smooth hand-trowel passes work best to smooth concrete.

greatly reduce the formation of prominent lines left behind by regular rectangular hand trowels.

Hand trowels are swiped over concrete in half-circle motions. Stretched out, this is a natural motion for your arm. Look at finished concrete slabs, like your garage, and you will notice definite trowel passes that were made in an arc shape. As you had to do with all of the rest of the finishing tools, you will have to raise the left-hand edge of your trowel when moving toward the left and the right side when going right. This prevents the leading edge from digging into the concrete surface. You will only make the mistake of not lifting a leading edge once, as the damage done by this will be more than obvious.

A rounded trowel motion is not mandatory for all finishing endeavors, however. If a slab you are finishing is quite small, like a stoop or step, you can simply apply a trowel on one side and maneuver straight across to the other (FIG. 11-9). This is especially true for slabs with unique designs or small width sections. When forming mowing strips or other extra narrow jobs, try to make them wide enough for a trowel to maneuver inside them (FIG. 11-10).

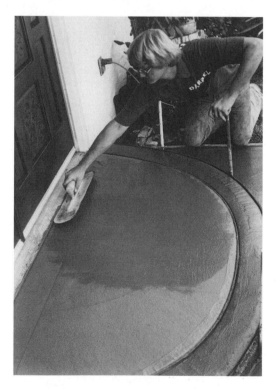

11-9 Finishing trowels are not always maneuvered in an arc motion. On small slabs, they can be operated straight from one side to the other. Here, a trowel edge is lightly placed next to the doorsill and then moved across the concrete toward a form.

As important as it is to raise the leading trowel edge, it is just as important to maintain trowels at workable angles. Raised too high, a trailing edge will scrape against concrete, causing tiny rock tips to pop up and cream to be scraped off. If the leading edge is too low, cream will swirl

11-10 When forming borders, mowing strips or any flat work design with tight dimensions, leave enough space to maneuver tools.

under trowels, leaving behind unacceptable rough finishes. Practice with hand trowels to get a feel for them, their correct application angle, and a hand position most comfortable and functional for the job.

Moving hand trowels too slow will not effect smooth finishes. Going too fast usually results in mistakes and blemishes. A steady controlled speed normally works out best. Move your arm about as fast as you would while operating a paint roller on a wall. You could go faster or slower but only practice can determine that.

Attempting to finish concrete that is still too wet will result in trowel lines and soupy ripples (FIG. 11-11). In these situations, use a fresno to see if its wide blade will do a better finishing job. Hand trowel work at this point is a waste of time. A wet slab will not hold a smooth finish. Conversely, if your trowel doesn't seem to do much to smooth a hardened surface section, you'll have to scrub up some cream and then apply a finishing swipe or two (FIG. 11-12). Use your smallest metal hand trowel for scrubbing. Vigorously rub it back and forth across the rough spot until a layer of cream is brought up (FIG. 11-13). The entire scrubbed area must have the same texture and swirl-type finish as the rest of the slab. Once this is accomplished, all holes and pockmarks should be filled.

Only very rough spots need to be scrubbed. Most other areas cream-up with just two or three vigorous trowel passes. Work small areas at a time with your scrubber trowel to bring up moisture and cream. Then,

11-11 Trying to finish concrete that is still too wet will result in lots of trowel lines and a soupy surface texture.

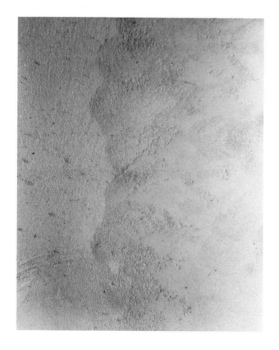

11-12 This is a rough concrete surface that needs to be scrubbed and then finished.

make a swipe or two with your larger trowel to finish the area off. Don't worry about very light, fine lines left behind by trowel edges. These will be wiped out by broom bristles later on.

Hand trowel finish work consists of two separate actions: scrubbing to bring up cream and moisture and finishing to smooth it all out (FIG. 11-14). One of the biggest mistakes novice hand trowel finishers make is

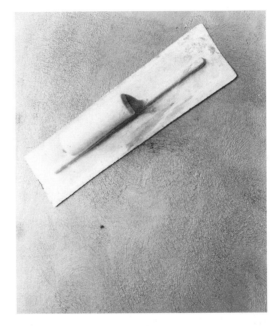

11-13 A rough but uniform texture shows that moisture and cream have been brought up to fill small holes.

11-14 This finisher reaches out to finish as much of the slab as can be reached from outside forms.

working concrete too much. Instead of just briskly working up some cream and then wiping it off, they continually scrub and finish. Scrub relatively smooth areas with three or four passes and then try to effect a final finish. Concerned that this effect will not be good enough, run a broom over it to really determine final effects. If it looks good, leave it and continue on.

Concrete work is no different than anything else when it comes to the old adage that practice makes perfect. If you are concerned about your concrete finishing ability, find a somewhat out-of-the-way section of concrete to practice on while it is still wet. Find a remote corner or section that will be covered with patio furniture or otherwise heavily broomed. Remember that finishing wet concrete results in a lot of lines left behind by trowel edges and, since wet mud is not hard enough, the finish will not be as expected. No matter. This practice helps you get a feel for the tool and to determine where your hand position on the tool works best.

Plan to finish center slab sections after successfully finishing around the perimeter outside of forms. Stretch out while leaning on a trowel, you might be able to reach much more of the surface than expected. This can really eliminate a lot of unnecessary knee board activity in the middle of slabs, while reducing your workload in the process. All areas inaccessible from the outside will have to be finished on knee boards.

11-15 This finisher is working backward on knee boards, finishing the concrete in front of him.

Use knee boards in a leap frog fashion working backward as you go. This way, you can trowel out impressions made by knee boards while progressing to another section (FIG. 11-15). The ultimate goal in finishing with knee boards is to finish the greatest amount of area while moving knee boards the fewest number of times. Reach out as far as practical from every side, because each time a knee board is placed on fresh concrete, some type of impression is made. Those impressions require scrubbing before finishing (FIG. 11-16). The more you move knee boards around, the more impressions you'll have to scrub out.

Along walls, especially rough textured ones, you might have to operate a trowel somewhat like a putty knife. Small sections of concrete might be very rough because a fresno could not adequately reach them. Use the tip of your trowel to initially pat concrete and then gently but firmly slide it away (FIG. 11-17). If more pressure is needed to wipe out extra rough spots, place the fingers of your free hand on top of the trowel tip and push down. Sometimes, a little concrete can be scraped off of walls and used to fill in holes and rough spots along them. Fill in as needed and then start the trowel swipe against the wall and move out. When you get toward the end of your reach, start lifting up on the trowel until it completely comes off the surface in one smooth motion; sort of like an airplane lifting off the ground.

Making hand trowel finishes absolutely perfect is only critical when finishing slabs that will not be broomed or otherwise custom finished, such as garage floors and tennis courts. Tennis courts are generally large pours that require a lot of forming and finishing expertise to perfect exact

11-16 Knee boards make an impression on concrete surfaces. Scrub knee board blemishes and then wipe them out with a finishing trowel.

11-17 Finishing concrete against walls might require extra trowel pressure. Push down on the end of a trowel with your free hand to help smooth rough spots.

grade standards and surface textures. Ordinary outdoor flat work requires a broom finish, at the least.

FINAL EDGING

You can go around slab perimeters with an edging tool after each fresno application but generally, you only need to after every other fresno (FIG. 11-18). Naturally, as concrete hardens you must maintain edges in excellent condition at all times, especially if fresno or trowel work results in lines or accumulations of cream.

Edgers must be kept flat on the surface in a side-to-side plane, raising the front edge to move forward and the rear edge to go backward. Too much pressure applied to the rounded side of the tool will cause edges to tilt down from form tops (FIG. 11-19). This results in the extreme outer perimeters of slabs to slope down slightly from the surface. On the other hand, too much pressure exerted on an edger's flat side edge causes deep lines (FIG. 11-20). For the most part, these lines are not major problems but simply create more work to finish a slab surface (FIG. 11-21). Therefore, it is a good idea to practice positioning and operating edgers so that the extreme outer edge does not slope or create deep lines. Although lines are wiped out with metal finishing trowels, the deeper and more frequent they appear, the greater chances are that they will turn into permanent surface creases.

11-18 Final edge finishing is done using edgers.

11-19 Excessive force on an edger's rounded side causes slab edges to slump down below form tops.

11-20 Too much pressure on an edger's flat side results in heavy lines that have to be wiped out with a finishing trowel.

Rough edges resulting from fresno work and moisture evaporation can be made crisp again with continued edger work. Try to minimize the amount of times you have to work edges to make them perfect, but do make as many passes as necessary to accomplish the job. This might sound contradictory, but the more you rub slab surfaces with finishing tools, the greater chances are of working up sand or grit. For rough edges, short back and forth strokes, steady and firm, will bring up cream and fill holes. When the area has been filled with cream, apply one or two light strokes to finish it off crisply and uniformly.

Most of the time, concrete finishers carry their edger tool with them while using hand trowels to finish perimeter slab areas. This way, they can finish edges and perimeter areas at the same time. Because trowel work can mark edges with lines or extra cream, do it before edging. After edges have been finished, apply a few light trowel strokes to wipe out inadvertent edger lines. By working on both perimeter finishing tasks and edges at the same time, finishers accomplish jobs quicker with less moving around. Conversely, should one helper be edging and another finishing, they might just get in each others' way.

Edging along stringers is just as important as edging around perimeter forms. For stringers in the middle of slabs, use a walking edger equipped

11-21 A rounded pool trowel works great for wiping out edging lines. Run a trowel lengthwise to knock down lines and then operate it from the form edge out to the finish area.

with extension poles. To effect final edges along stringers, bring your hand edger along while leapfrogging around the slab with knee boards and hand finishing trowels. As you reach stringer sections, edge along them as you would outer forms (FIG. 11-22).

Edging work must also be completed along expansion joints located in the middle of slabs, whether they are felt, stringer, or benderboard. Use a walking edger, as needed. Be sure to fill in holes along edges and finish areas next to joints with a hand trowel (FIG. 11-23). Apply the edger tool next to expansion joint edges to make them crisp and perfect (FIG. 11-24). Wipe out any edger lines and then finish surface areas with your trowel as far out as you can reach. By doing both finishing and edging chores at the same time, you accomplish all of the necessary work in that area while only having to place and work off of one knee board at a time.

Notice that in FIG. 11-24, the finisher edged both sides of the expansion joint at one time while positioned on just one side of it. This eliminates the need to edge opposite expansion joint sides when finishing concrete on that other side. By edging both sides of expansion joints at the same time, you are guaranteed that concrete will be the same consistency and hardness for both edges.

Final seam finishing is done much the same way as edges. After a final fresno application, run your seamer tool along grooves to effect crisp seams, including the base of grooves as well as the sides. Don't worry about lines made by the tool's outer edge; they'll be wiped out with a

11-22   Each section of the slab should be completely finished while you are there. Bring an edger or seamer to finish around obstacles, as is being done along this stringer.

11-23   A benderboard expansion joint has nearly been covered with cream from repeated fresno applications. This finisher will use a hand-held edger to complete work on the joint after trowel finishing surrounding concrete within reach.

11-24 Both sides of this benderboard expansion joint have been edged from the finisher's location.

hand trowel. Work seamers back and forth along rough spots. You should not have to use a 2-×-4 guide, a seam's groove should be stable and supportive. If you experience trouble operating a walking seamer to smooth rough surfaces, take extension poles off and use it as a hand seamer. Simply finish seams as you did with edgers—out on the slab with knee boards.

BROOM FINISHING

The evolution of broom finishes was a concrete finisher's dream come true. Besides adding required traction to slabs, broom lines cover a lot of minor surface finish mistakes like slight trowel lines and some very light rough spots. Because broom bristles actually penetrate concrete surfaces, albeit not very deep, they are able to essentially roughen finishes to where small amounts of cream are able to fill in tiny holes and crevices (FIG. 11-25).

Walkways are generally broomed across their width for maximum traction capability. Patio slabs are broomed in line with their slope, generally in a direction perpendicular to houses or other structure walls they were poured against. Lines going in the direction of the slope on a patio slab, for instance, helps water rinse better because it flows toward the slab's lowest spot. If applied sideways, these lines would tend to dam water and possibly trap pieces of dirt and debris, making rinsing a bit more difficult and time-consuming.

11-25 A broom has been applied to the left side of this slab, resulting in a uniform finish of light penetrating lines.

The wetter concrete is, the deeper and more definitive broom lines will be. If your slab is a driveway on a very steep grade, for instance, you could apply a stiff-bristled broom to the concrete surface soon after fresno and edger work to effect an exceptionally rough finish (FIG. 11-26). This finish would greatly add traction for vehicles using a driveway in wet weather.

Concrete that has set up and is hard, might not accept much bristle penetration at all. In these cases, you might have to push and pull a soft bristled broom across the surface repeatedly to accomplish lines. In extreme situations, finishers have dunked broom bristles in buckets of water and then applied them wet.

Most slabs that have been finished on schedule will have enough moisture on their surface to allow neat, crisp, and sufficiently deep broom lines. Finishers can simply place their broom at the far end of a slab and then lightly pull it across the surface one time in a steady motion. Pushing a broom across slabs with accommodating surface texture (moisture) will leave lines that are far too deep. It will also amass cream build-up on the leading broom edge, causing blotches and wide scratches instead of thin clean lines.

In order to lift and move a broom head from outside formed areas to far reaching slab sides, you will have to attach extension poles to your broom handle. This is easily done by slipping an extension over the broom handle and then securing it with a wood screw inserted through an open hole on the extension and into the broom's wooden handle. If you do not have a soft-bristled push broom, regular concrete brooms are located at most tool rental facilities.

11-26 A broom was applied to this step while concrete was very wet, leaving behind deep heavy lines for extra traction when the surface is wet. An edger was used after brooming to smooth the perimeter and form a border around the center.

Pulling a broom across a concrete slab takes no special talent. All you do is raise the broom in the air, stretch it across the slab to its furthest point, gently lay it down and then pull it straight back toward you. If you see that bristle impressions are penetrating too deep, simply apply upward pressure on the poles to reduce the broom's weight on the slab. This will lessen its ability to dig into the surface and result in lighter lines (FIG. 11-27).

If you want all broom lines to run in the same direction, the most common pattern, you'll have to start at one end of the slab and work yourself across it, one broom length at a time. For a patio slab poured against the back of a house, for example, you would stand at a corner of a side form and back form edge while facing the house and lay down a broom line along the side edge; your broom will span from the side form in toward the slab. Once that pass is made, you simply lift and place the broom for another pass right next to the first, and so on until the entire slab is broomed.

If concrete is a little hard and one pass doesn't quite give the line depth desired, you can overlap each pass by half a broom length. Each pass will actually receive two broomings. Because of the span, your broom head operates at an angle as opposed to a flat plane in which brooms are normally operated for sweeping. This is of no consequence and will not cause any problems as long as it is pulled straight.

Other broom line designs or patterns can be applied. By wriggling a

11-27 To make straight lines, a broom must be pulled straight back. If lines appear to be penetrating too deep, simply apply upward pressure on the handle to reduce bristle weight on the surface. If lines are too light, apply downward pressure on the handle.

broom back and forth in a sideways motion, you can put on a wavy broom finish. As you pull the broom back with one hand, have the other positioned further up the handle closer to the broom head. Work the broom back and forth to create the pattern desired. Be sure to maintain the same sideways movement throughout each pass so that all passes will be symmetrical.

Another way to dress up broom finishes is to run an edger tool over all the edges after brooming. An edger will wipe out broom lines and leave a smooth border around a broomed slab. Before doing this, you must be certain that the smooth border will not hamper your ability to walk with good traction on slabs or walkways when they are wet.

In tight spots around slabs where a broom is impossible or exceptionally difficult to operate, simply remove the handle and use the broom head only (FIG. 11-28). In addition, you could just use a smaller foxtail broom.

A "windshield wiper" broom finish pattern is unique. It is applied with a fox tail broom that is moved over slabs in short, arm-length semicircles about 2 to 3 feet in diameter. To do this, you'll have to carry a foxtail broom with you as you are out on a slab with knee boards and trowels. After a section of concrete has been trowel finished, simply extend the broom out as far as your trowels were and, without moving your arm, bend your wrist from side to side while applying the broom. Pull the broom straight back while doing each section. The end result will

11-28 Broom applications in tight areas and close to form edges are not always easy with extension poles attached to brooms. For best access, remove poles and use brooms with their regular handle or without any handle at all, as is being done here to touch up a blemished area.

be a broomed slab consisting of semicircles all running in the same direction, with equal width and equal broom arc.

No matter what kind of broom finish design you apply, lines that are too deep can be troweled out and done over again later, when concrete is harder. If you notice this problem on your first broom pass, stop. Don't broom an entire slab that is too wet and then decide to trowel it all out and broom it again later. That is too much work. As soon as you notice that concrete is not quite ready for brooming, stop the process, trowel out what you have already done, and then wait until the surface has set up enough to accept the type of broom finish desired.

On the same note, if broom lines are too light, put downward pressure on the broom handle to help bristles penetrate deeper. If that doesn't work, pull and push the broom across each section a number of times using downward pressure as needed. As a last resort for slabs that are still too hard for ordinary brooming, dip broom bristles in a bucket of water and apply them soaking wet. Shake out excess water so the broom is not dripping continuous lines of water.

Chapter **12**

The last hour

Although there is still some work left to be done, concrete pouring and finishing should be complete by the time you reach this phase of the operation. Congratulations! Now, you can really sit down for a few minutes to catch your breath, have a cold drink, and admire your work. If you have followed directions up to this point, concrete should be completely finished and broomed, excess or spilled concrete cleaned up, and the tops of forms scraped off so only slight residual mud is left. Tools should be in good shape because they were rinsed with water after each use.

After your short rest, though, be prepared for some additional tasks. The last hour after concrete is trowel finished is filled with a number of small chores designed to protect new slabs and essentially complete jobs that leave work sites clean, tidy, and looking good.

PROTECTING CONCRETE

The longer concrete ages, the harder it gets. In fact, concrete takes about 99 years to reach its maximum strength and hardness. Rest assured, however, because concrete quickly reaches most of its strength (almost 99 percent) in only about two weeks. Along with short-term barrier protection to keep people and pets from walking on new slabs, you must consider how you will maintain moisture inside slabs for a couple of weeks. This time segment is about how long it takes for concrete to dry out and reach a hardened state where normal activities will not permanently scratch or mar surfaces. During this two-week "incubation" period, slabs are normally susceptible to cracks, scratches, and other damage, such as erosion from water falling off roof eaves.

Typically, cracks occur on fresh slabs because concrete dried out too fast. Mixtures need to cure rather slowly to allow all of the internal elements time to fully set up. This is evidenced by concrete's color changes. Fresh concrete is generally a dark color; professional finishers call it green. Cured concrete is white and is a definite sign that moisture has just about completely evaporated from green concrete. It takes about two weeks for new slabs to turn white and you should not walk on or use slabs until they have turned a uniform white color.

One way to help concrete cure slowly and maintain a sufficient moisture base is to keep it wet with water spray (FIG. 12-1). For two weeks, sprinkle slabs with a water spray at least six times a day. Ideally, they should be wet down every time any section starts to dry out; about once every two hours during the day. Concrete will cure under water, and you could even build a dam around slabs to maintain an inch or two of water on top of them. Although some special concrete jobs might need to be protected this way, building dams around slabs is quite impractical. Water spray from a garden hose is plenty good enough.

12-1   New concrete slabs should not be allowed to cure too fast, especially during warm summer weather.

Protecting slabs with water spray has a couple of minor drawbacks. One, of course, is the time required to actually do it. Someone must be available and has to remember to wet the slab down every two hours or so. The other drawback is most prevalent during warm summer weather. As the sun beats down on a slab, its surface temperature rises. Hitting it with a spray of cool water causes a thermo-shock, making the finish rapidly contract after expanding from the heat. This drastic change in surface temperature can cause "map cracks," a series of very fine, hairline surface cracks that wander all over finishes.

Map cracks are very shallow and do not pose strength-related problems, they are only cosmetic in nature. To avoid map cracks, make sure slabs are wet down consistently. Keep them wet and cool so that their surface temperature remains relatively stable.

Because professional concrete finishers cannot remain at job sites for two weeks to continually wet down slabs, they rely on special concrete cure and seal products to help slabs maintain required moisture (FIG. 12-2). This material is sprayed on finished slabs to seal surfaces so moisture does not rapidly evaporate. It is easily applied through a common variety garden sprayer. Concrete sealer products generally sell for around $12 a gallon with a gallon able to cover roughly 250 square feet.

12-2 A concrete sealer product helps slabs maintain sufficient moisture to ensure proper concrete curing. Here, spray is applied through a garden sprayer.

Concrete sealers are available at some lumberyards and hardware stores. Most often, though, sealers are acquired through ready mix concrete plants. The concrete truck driver might be able to bring it along to your job or you can go directly to the plant to buy it. Ask if you should bring along some plastic milk containers or other gallon jugs. Some concrete plants purchase this product in bulk supplies and will fill your containers from a 55-gallon drum. Because various sealer brands have different mixing or application ratios, you should consult the concrete plant dispatcher for specific application instructions.

Another way to help keep fresh slabs protected is with heavy-duty plastic tarps. Once concrete is finished and set up, lay a sheet of black plastic on top of it to help keep moisture from evaporating (FIG. 12-3). This may be one of the least expensive and care-free methods of allowing concrete to cure at its own rate. Be sure concrete is hard enough to accept a layer of plastic before putting it on, however. Wet concrete finishes can be scratched or gouged by wrinkles in plastic. Test finishes for hardness by touching them with your finger. If an imprint is made with your finger, a crease or wrinkle on plastic will do the same thing. If there is no noticeable impression, then plastic should not make any marks either. Secure plastic in place by laying 2×4s or bricks on top of it around the outside perimeter of the job, not on top of the concrete.

12-3 Plastic sheets protect new concrete from fall and winter rainfall and can be used in conjunction with concrete sealer during extra hot weather to prevent excessive moisture evaporation.

Plastic tarps are essential during winter—not so much to keep moisture in concrete, but to keep rain and cold away. Even semi-cured slabs can be ruined by raindrops and water runoff from roof eaves. Once concrete has sufficiently hardened, cover it with plastic stretched tight so wrinkles and creases are not allowed to rest on the finished surface. Plastic will lessen the impact of water drops and protect slabs against surface erosion.

Because concrete contains moisture, slabs must be protected against freezing temperatures. If moisture inside fresh concrete is allowed to freeze—essentially turning into ice—it will expand, causing cracks and might even result in chunks of the slab surface popping out. It would look like someone smacked different parts of the surface with a hammer and chipped out pockets of the finish. If possible, delay concrete pours until spring when weather has warmed to above freezing.

Professional concrete finishers protect slabs in freezing weather with plastic and straw. Once a slab has hardened enough to support a plastic cover, a thick layer of straw is spread over it to provide insulation. Plastic helps to keep concrete's own heat and moisture confined to the slab and a few inches of straw insulates against outside temperatures. In addition, a very slight amount of heat generated by straw's decomposition also helps to keep concrete from freezing.

CLEANUP

There is no getting around the fact that concrete work is messy. No matter how hard you try, mud will find its way to forms, tools, walls, preexisting slabs, and just about everything else located close to your job site. Cleanup should begin as soon as practical because fresh concrete is much easier to clean off than that which has sat around for a day or two.

Tools and equipment must have first cleanup priority. Unless they are cleaned promptly, concrete will set up on them making eventual cleaning a rigorous task or will simply render them useless for future jobs. Use a five-gallon bucket filled partially with water to clean finishing tools after each use. When the finishing job is complete, thoroughly clean tools with fresh water and a brush. Try to remove every spot of concrete, especially on blade faces and around bull float and fresno swivels.

Forms and stakes must also be cleaned. Water and a brush will make quick work of cleaning concrete off of them so that they are usable for future concrete jobs or other things, such as landscaping borders, patio covers, and the like. Be sure all nails are pulled out. To make this job easier, finishers fill a wheelbarrow full of water and put stakes in it. Once washed, they are neatly stacked and put away.

Areas around a job site need cleaning attention as well. The concrete truck driver had to clean out the chutes and you will have to remove whatever concrete residue resulted from that. For the most part, chutes are usually rinsed in the street, close to a curb. Use a square-point shovel to scoop up sand, aggregate, and cream, then wash away the rest with plenty of water. Although water pressure is generally enough to rinse away most concrete residue, you might need to use a stiff-bristled broom or brush to loosen stubborn accumulations.

Obstacles located in the middle of slabs also need to be cleaned before concrete has the opportunity to permanently set up on them. An old trowel or putty knife works good to scrape up loose concrete, while a wire brush does an excellent job of removing everything else. Leftover sand and crumbs of hardened concrete can be swept up with a whisk broom and dust pan.

Fresh concrete that is allowed to cure on top of old concrete will soon become a permanent fixture. Plan to clean preexisting concrete slabs as you would other obstacles. Use an old trowel or putty knife for the big stuff and then finish the job with a wire brush. You should do this right away. Waiting a day or two will allow new concrete to harden and make it very difficult to break loose and remove.

While using water to clean up areas in front of your house, be sure to control water streams so that they are not directed on top of new concrete. As accumulations of sand and aggregate are amassed, use a shovel to scoop them up into a wheelbarrow for proper disposal.

Heavy accumulations of excess concrete materials must be properly picked up and removed to an adequate disposal site. Use a wheelbarrow or five-gallon bucket, whichever is most practical, to transport sand, aggregate, and cement residue to a dumping site in your backyard or other suitable place (FIG. 12-4). If you have to take this debris to a landfill, put a piece of plastic down in the bed of your pickup truck or trailer before putting concrete residue in. This will help keep your truck or trailer bed clean.

12-4 Accumulations of concrete can be scooped up easily with a square-point shovel.

No matter where concrete truck chutes were rinsed or concrete overflow spilled, you can clean the area with water to make it look just like it did before the job was poured. All it takes is a little time and effort. The more neat and tidy job sites look, the better your concrete work will also look (FIG. 12-5).

PULLING FORMS

Forms and stakes are generally pulled the same day concrete is poured, provided concrete has set up to the point where it can support knee

12-5 A few minutes of conscientious cleaning goes a long way toward a professional, crisp, and eye-appealing job well done.

boards without them making impressions on the surface. However, there are exceptions! Do not pull forms on the same day slabs were poured during cool, damp winter or autumn days and never pull forms on the same day exposed aggregate finishes were implemented (a custom concrete finish).

In these conditions, concrete has only had enough time to set up sufficiently on the surface. Underlying concrete might still be very wet and unstable. Pulling forms too early during cool, wet weather and on slabs exposed to a lot of water for exposed aggregate finishes, might allow unstable concrete to pull loose or slump down, ruining an otherwise perfect job. In these situations, plan to wait at least 2 or 3 days before pulling forms.

Pulling stakes and forms entails nothing more than extracting nails from stakes, loosening and removing them from the ground and then pulling forms horizontally away from slabs until they are free. Use a claw hammer or nail puller to take out all nails. Hit stakes on their sides with a small sledgehammer so they move back and forth along the sides of forms to loosen them. Pull stakes straight up to extract them from the ground. Stakes that have been driven a long way into the ground might require repeated hits with a hammer and some extra strong pulling to get them out.

Once all of the stakes have been removed, you can start to gently remove forms. Never pull forms straight up. Instead, always pull them horizontally away from slabs first. Often, forms are sort of attached to concrete by way of suction. Pulling straight up on forms can cause chunks of the concrete surface to come up with them. Although these problems can be easily repaired, it is really best to avoid them. By pulling forms away horizontally, you greatly reduce the chances of breaking loose chunks of surface concrete and forms will also come off much cleaner.

Start at one form end and gently break it loose from concrete by softly tapping on the board with a hammer horizontally. As you see this form start to separate from concrete, gently but firmly ease the form out and away. You might have to wriggle a form back and forth horizontally until the rest of its length breaks loose. Once it is free, pick it up and take it to the cleaning area. Be careful while carrying long forms that they are not brushed against the slab or knocked against anything that might break or be damaged. Clean forms with water and a brush.

REPAIRING EDGES

Sometimes, no matter how careful you are, a chunk of fresh concrete will come off of an edge as forms are pulled. This might be caused by a sliver of wood that has had concrete poured around it or from other unavoidable factors. Fortunately, this problem is easily corrected. When forms are pulled the same day as the pour, concrete is still green enough to allow edge repairs. All you'll need is a hand edger and a 2-×-4 block about a foot long, or long enough to extend past the damaged edge on both sides.

Quite simply, just put the 2-×-4 block in place next to the damaged edge so that it extends past it on both ends. It will sit just like the original form did. Then, scrape up the chunk of concrete that was pulled off or an amount of cream from the original form equal to what is missing. You might have to roll this wad of mud in your hand and add a few drops of water to it in order to get it into a creamy and usable consistency. Once that's done, put the wad of mud on top of the damaged edge and then operate your hand edger back and forth over it until it blends in and looks the same as the rest of the slab.

The wood block will support your effort so that you can maneuver a hand edger as you did with the original form in place. Remember to apply pressure against the 2-×-4 block with your free hand or knee because there will be no stakes there to support it.

If you pull forms a few days after a pour, edge repairs can still be done but will require a different technique. Because concrete will have hardened, you will need to coat the damaged area with concrete glue (Elmer's white glue will also work), and then gently replace the chunk that broke off. If the chunk just crumbled and is in no condition to be reattached, you might have to buy a sack of cement or concrete patch material and mix up enough material to make repairs. Be sure to follow label instructions on the package to ensure suitable repair results. Concrete patching products are sold at lumberyards, hardware stores, and through some ready-mix concrete companies.

OVERVIEW

Once all of your metal concrete tools have been washed and dried, you should consider coating them with some type of a rust-prevention material. Many finishers have had good results spraying their tools with a commercial lubricant, such as WD-40. A coating of this or a similar product prevents rust from forming on fresno and trowel blades and keep swivels and connectors in good working condition. Wood floats are sanded to remove rough textures and/or slivers.

Although professional concrete finishers take great pains to prepare and pour jobs perfectly, they sometimes have to contend with unpredictable cracks. Hog wire and/or rebar reinforcements help to prevent cracks from enlarging and separating, but you will need to apply any of a number of concrete patch products to seal cracks and prevent water from penetrating the voids created by them. Be sure to read product labels to ensure you purchase materials designed for the repairs you have in mind.

Chapter **13**

Custom concrete ideas and large pours

A lot of do-it-yourself homeowners are so eager to pour concrete slabs and walkways over bare dirt to enhance their yards, they fail to consider that continuous-plain concrete can sometimes make landscape designs appear rather sterile. To some degree, planters along walkways and formed stringers inside slabs make otherwise plain concrete flat work designs look more custom. But to really turn ordinary concrete flat work into custom creations, you have to consider a variety of special options.

Along with custom finishes, there are a number of important factors that must be considered when pouring large concrete jobs, especially those that will require more than one concrete delivery truck. There are also occasions when large amounts of concrete must be delivered to job sites where trucks cannot get close to forms. In those cases, you'll have to have a concrete pump on site to deliver concrete from a truck to forms through special, heavy-duty concrete hoses.

Custom finishes are usually applied to slabs after concrete has set up and been finished. Some special finishes are real easy to apply, while others require plenty of forethought and a definitive work plan. While initially designing your concrete flat work needs, consider what you might do to make the job a truly custom one. No doubt a custom job requires extra work, but you'll only have to do it once and it will last virtually forever.

COLOR

A wide range of color dyes are available for customizing concrete. Powdered dye can be mixed into concrete at batch plants so that entire loads are the same color. You can also purchase dye and sprinkle it on wet concrete surfaces and then float it in with a bull float. Color charts and cost figures are available at concrete batch plants. A concrete company's dispatcher can help you with color selection, the amount needed, and cost considerations.

Adding color to concrete can be quite expensive. Each dye is different and so is its cost. Expect to pay anywhere from $8 to $50 extra per yard for color added to full loads from a concrete company. Applying dye dust yourself is much less expensive simply because you will use significantly less. Concrete color dye is generally sold by the pound, with one pound covering about 20 square feet when dusted on and floated in with a bull float (FIG. 13-1). Prices start at about $3 per pound and go up.

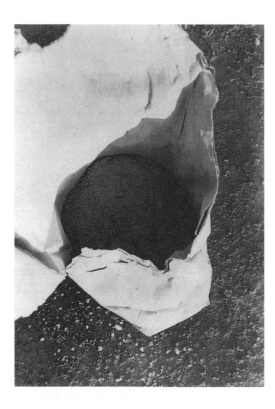

13-1 Concrete color dye is a powder material sold in bulk to ready-mix concrete companies.

Applying concrete dye is a time-consuming and messy chore. Concrete dyes are very potent and stain just about everything they come in contact with. One advantage of having your entire concrete load color dyed from a concrete batch plant is that entire slabs will be an identical color all the way through. Should a corner crack off or a major chip develop, color will still be there. Dusted on dye only penetrates surfaces about 1/4 to 1/2 inch, and should a chunk break off of a corner or chip occur in the center, concrete gray will be exposed. In addition, dyed concrete delivered as an entire load is thoroughly mixed and enhances the chance of having an identical tint over an entire surface. Spreading color dust by hand, especially by novices, could be done unevenly, resulting in blotchy or irregularly tinted finishes.

Colored concrete must be sealed. This is done once the slab has hardened enough to allow gentle walking on it without causing impressions on the surface. A liquid sealer is applied through an ordinary garden sprayer, just as the sealer used for regular concrete is. If colored concrete is not sealed, it has a greater chance of fading after continued exposure to sunlight and weather. Some colors are more susceptible to fading problems than others. Your concrete dispatcher can explain the various complexities of each color, how they are affected by your region's climate, and which sealer is best to use.

ROCK SALT FINISH

A rock salt finish features hundreds of small shallow holes all over a slab's surface (FIG. 13-2). After slabs have been completely finished and broomed, finishers toss handfuls of rock salt over the surface. Then, they get out on the slab with knee boards and pound salt granules into the slab with a hand float or trowel. Several days later, the salt is dissolved by continued water spray or actual rinsing. The final effect is a uniquely textured and designed hole pattern (FIG. 13-3).

13-2 This custom patio has a series of sectional seams, open planters, and a rock salt finish.

13-3 This brick insert design looks great and is enhanced with a rock salt finish and seams.

Rock salt is available at most grocery stores and supermarkets. It comes in different sizes ranging from about 1/2-inch-diameter rocks to granules a little smaller than garden peas (FIG. 13-4). Extra large rock salt granules are generally too big for concrete flat work. For best results, use granules about the size of a pea or smaller. The smaller size is easier to pound into slab surfaces and makes designs look more uniform and crisp.

The most important factor to remember when creating rock salt finishes is to apply salt granules as uniformly as possible. Too many spots heavily covered in salt and others just lightly covered results in an uneven and awkward-looking texture. Spread salt thin at first and then add as necessary to evenly average salt applications.

Because rock salt finishes are broomed first, there should be plenty of traction available when slabs are wet. About the only drawback to rock salt finishes is that holes can become places where dirt and debris accumulate. Simple sweeping is generally not enough to clean them. They have to be rinsed with water from a garden hose (FIG. 13-5).

EXPOSED AGGREGATE

The term *exposed aggregate* means just exactly what it says. Once a slab has been finished with hand trowels and edgers, the very top layer of cream is washed off to expose the aggregate beneath (FIG. 13-6). When done correctly, it looks quite custom and generally blends well with all sorts of landscape designs (FIG. 13-7).

13-4   These custom rock salt-finished steps extend out the side.

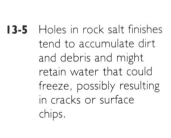

13-5   Holes in rock salt finishes tend to accumulate dirt and debris and might retain water that could freeze, possibly resulting in cracks or surface chips.

13-6 This exposed aggregate slab is enhanced with brick inserts and a handsome wood deck bridge over a small pool. The exposed aggregate finish blends exceptionally well with rocks around the pool and the walkway's border.

13-7 The exposed aggregate finish on this walkway and front porch patio could have been rinsed again and then sealed with a concrete sealer to preserve a darker, richer color.

Because of its rough texture, an exposed aggregate finish is difficult to sweep clean. Like a rock salt finish, expect to rinse it off with water to remove accumulations of dust, dirt, and landscape debris. In addition, because aggregate is exposed on the surface, it is hard on bare feet and especially unforgiving on knees and elbows of youngsters who might fall down on it. One way to eliminate some of the rather rough texture problems associated with exposed aggregate finishes is to order smooth rock aggregate in your concrete mix. Some concrete companies get their aggregate from river bottoms where gravel has been worn smooth. This rock is separated by size, and large aggregate is fed through mechanical crushers where it is broken down into 1 inch and smaller diameters.

Depending on your location, smooth rock might be readily available or might have to be special ordered. Check with your concrete dispatcher for details on availability and cost considerations.

Aggregate comes in different sizes (FIG. 13-8). Most finishers work with ³/₄ minus, which means the largest aggregate size is ³/₄ inch in diameter with some smaller material mixed in. A full load of pea gravel aggregate used for an exposed finish has very small rock measuring about ³/₈ inch in diameter and smaller. The choice is yours, and is generally based on which aggregate is most eye-appealing to you.

Producing an exposed aggregate finish requires washing off the very top layer of concrete cream from an entire slab surface. This is a very messy job. Be prepared to clean up concrete residue from around forms, in flower beds, and anywhere else it washes into.

13-8 Properly rinsed and then coated with a concrete sealer, these two samples look great. Smooth, 3/4-inch-minus aggregate was used for the pour on the left and pea gravel on the right.

ETCHED FINISHES

One of the newer types of custom concrete finishes developed by innovative professionals in recent years is called the etched finish (FIG. 13-9). This custom effect is easy to apply and sometimes fun to do. There are no set patterns. You just etch along all forms and round off all corners. The rest is all done free hand.

13-9 This etching finish was applied freehand. A 1-inch-wide wire brush was used to rough out lines and make the steps look almost as if they were made out of separate stepping stones.

Concrete is first trowel finished and broomed, just like an ordinary slab. A one-inch-wide wire brush is then rubbed along all perimeter edges, expansion joints, stringers, and seams, with all corners rounded off in the process. A wire brush does not have to penetrate very deep—only enough to roughen smooth finishes and leave behind a solid etched line. You could try to etch deeper and expose aggregate but run the risk of etching so deep that definitive ridges result, creating tripping hazards.

After all of the straight borders have been etched and their corners rounded off, you can then apply your design to the rest of the slab free hand (FIG. 13-10). The resulting design resembles a flagstone deck, with etched lines similar to mortar between steps. Because you will have actually etched the concrete surface and roughened its texture, lines will remain darker than the rest of your slab. This contrast enhances the slab's custom appearance.

INSERTS

A lot of concrete finishers use brick borders and dividers to liven up slabs (FIG. 13-11). Bricks come in a wide variety of colors, styles, and shapes—almost any combination can liven up a plain walkway. Walkways can be

13-10 A freehand etching finish applied with wavy lines for a unique appearance. All intersecting lines and square corners were rounded off.

13-11 Brick inserts in this patio slab blend with other brickwork around the raised fireplace area and planters.

13-12   After this walkway was poured and finished, a brick border was installed around it.

bordered with red brick to extend their width or to match other brick work in the landscape or house (FIG. 13-12). Patios can be bordered and uniformly divided with brick to create a warmer atmosphere or to set off certain areas for barbecues or fire rings (FIG. 13-13). Creative finishers have even gone so far as to implement special brick designs in their slabs for truly custom effects (FIG. 13-14).

Bricks are inserted into slab areas after concrete has been poured and cured. The slabs are formed in such a way that areas set aside for bricks are left open. Form installers must not only leave enough room for bricks, but must allow room for mortar as well. Once a concrete slab has cured, about two weeks, bricks are placed in position and secured with mortar (FIG. 13-15). Be certain that areas left open for them are the right size, and don't forget to figure in the width of forms when allotting open space for bricks and mortar.

DECORATIVE SEAMS

Similar in scope to stringers, decorative seams break up solid concrete slabs to offer unique patterns and shapes (FIG. 13-16). They are applied with a seamer operated next to a 2-×-4 straightedge guide. Seams can be inserted at angles to each other or matched in separate patterns to create divisional slab sections (FIG. 13-17).

Incorporating a system of symmetrical seams in your slab, along with a custom finish and planters, can add a great deal of visual appeal to an

13-13 Brick inserts are in line with patio cover support posts.

13-14 Seams come off of every featured corner of this slab. Otherwise, cracks would eventually appear.

13-15   A breezeway connects this house with a detached patio. Brick inserts blend well with the brick wall in the patio and help to break up otherwise solid concrete.

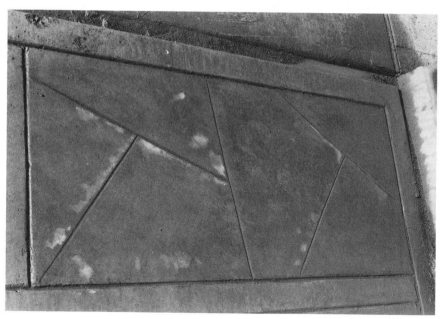

13-16   Decorative seams can be installed in any pattern. A unique set of angled seams applied to concrete that is also bordered by seams makes for an interesting walkway.

13-17 A decorative seam design divides this slab into equally sized sections for a uniform appearance.

otherwise plain patio. There will, of course, be more work involved in the process but the creative effects are an enduring landscape feature.

Seams must initially be installed right after concrete is bull floated. While still relatively wet, concrete will easily accept seamers and allow aggregate to be pushed out of the way. Waiting too long to apply a seamer will find concrete stiff and aggregate very difficult or impossible to push down.

If your slab is large and you want to decorate it with a series of seams, be sure to secure a walking seamer. This will allow you to effect seams while standing outside of forms. Seam grooves can be perfected later when you are out on the slab with knee boards, trowels, and a seamer (FIG. 13-18).

SPECIAL BORDERS

Just as concrete slabs can be bordered by bricks, tiles, and whatever else you might have in mind, concrete can also be poured to outline (border) brick patios, flagstone walkways, and just about anything else you find attractive. With a bit of inventiveness, you could create a border around slab perimeters with just an edger and seamer.

After an initial edge has been run along forms, try running an edger with its rounded side positioned in the concrete while the straight side follows forms—use it reversed. This results in the rounded side creating a deep cut in the concrete surface a few inches away and perfectly in line

13-18 This seam design offsets the open planters.

with forms. By following that deep cut with a seamer tool, you can create a border that looks like it is a separate, outlining curb (FIG. 13-19).

Borders of all types have been used to dress up otherwise sterile white concrete. Use your imagination and plan for concrete work to blend with its surroundings. For example, landscapes with lots of wood

13-19 Operating an edger with the arc side reversed creates a line a few inches in from form edges. Follow that line with a seamer to create a border around slabs.

timber planters are attractive when slabs and walkways are designed with pressure-treated 2×4s, which are then left in place after the pour. This additional wood helps concrete flat work better blend with the landscape.

For asphalt driveways or parking spaces that suffer continual broken ends, consider adding a concrete ramp. Asphalt butted next to concrete is much less susceptible to damage. First, square off asphalt by cutting it straight with a concrete cutter. Then, simply pour concrete against this end as if it were a form and finish it like any normal slab. Be sure to put broom lines on it to ensure adequate vehicle traction. This ramp will not only help to prevent asphalt damage, it will add a crisp and clean look to your driveway apron or parking space.

LARGE POUR CONSIDERATIONS

Large concrete pours can be anything over a yard for complete novices or jobs requiring up to 10 yards or more for more experienced finishers. As a general rule of thumb, however, let's say that any flat work job over 3 yards is going to tax someone who has never done any concrete work before, especially if he or she does not have an experienced helper available to assist.

It is a given that all concrete placement and finishing tools have to be on site before a concrete truck arrives and that forms and stakes have to be set correctly. In addition, you must have enough helpers on hand when pouring concrete. This is especially critical when pouring large jobs. Having enough help on hand cannot be stressed enough. You will need wheelers, screeders, screed helpers, a tamper, someone to run a bull float, someone to fill in and help with this and that and the other thing, and on and on. The more help you have, the better.

Large pours of more than one truck load will require that the truck go back to the plant and fill up a second time. In lieu of waiting for just one truck to come and go, try to schedule the first truck at a specific time and a second separate truck with the cleanup load 45 minutes later. You must be absolutely certain how much concrete is needed on the second load, however. If, for some reason, the second truck brings out 2 yards and you really needed 3, that extra yard could easily cost you $100 in cleanup fees. This is because the truck could have brought all of the concrete you needed once but now has to make another trip back to the batch plant and then all the way back out to your job site. This creates extra overhead costs in fuel, driver time, and missed deliveries for other customers.

On special jobs with lots of unusual angles, curves, or steps, it is always a good idea to order an extra 1/4 yard of concrete. The cost is minimal when compared to cleanup fees and it is very difficult to accurately calculate yardage for unusually shaped jobs.

When pouring a large job, always be certain measurements are correct and depth throughout the entire area is uniform. Should you have trouble calculating correct yardage for odd-shaped jobs, ask the dispatcher if there is anyone available that can come out to your job site and estimate the amount of concrete needed. Concrete companies sometimes employ salespeople who will gladly come out to assist you in estimating yardage.

CONCRETE PUMPS

Concrete pumps are machines that literally pump concrete from a hopper through hoses to formed areas. They are absolutely indispensable when it comes to delivering large amounts of concrete to job sites where wheelbarrows would have a tough time maneuvering, like uphill or a 100 feet or more from a delivery truck. Independent concrete pump owners and some concrete companies operate these pumps, and you need to coordinate their arrival with the delivery times set up with the concrete company. They can be found in the yellow pages of your telephone book under ''Concrete Pumping.''

Two basic types of concrete pumps are generally available. The smallest type is called a line pump. It looks like a large cement mixer and is towed behind a pickup truck or other work vehicle. It pumps concrete through a small hose stretched from the pump to the work site. Because of its small size, line pumps can only handle aggregate the size of pea gravel. Originally designed to pump grout into retaining walls, innovative masonry workers and flat work finishers soon realized that it could also be used to pump concrete, especially for jobs over 5 yards where a truck cannot get close to forms.

The other type of concrete pump is a very large machine mounted on a big truck equipped with an articulating boom. This rig is plenty big enough to angle up and over houses and other structures to reach forms. Its size can easily handle regular concrete mixes of $3/4$-minus aggregate, sometimes even 1-inch aggregate. This type of pump is perfect for large pours of 10 yards or more because there is no hose to drag around. The boom is simply maneuvered to accommodate concrete placement.

Concrete pump fees vary a great deal from one area to another. You will have to call a few pump companies to determine basic fees and schedule availability. Line pumps are usually less expensive, ranging anywhere from a basic rate of $70 plus $1.50 per yard and up. Large boom pumps are considerably more, generally starting in the $120-plus range for initial setup plus an extra per-yard charge.

The extra money it costs for a pump truck is greatly offset by the amount of wheelbarrow work saved, especially for pours in excess of 5 to 7 yards where wheelbarrows would have to be maneuvered more than 75 to 100 feet from a truck. If in doubt about whether you need a pump, ask the dispatcher for advice. He or she should be able to give you the criterion used at their facility to determine when pumps are most advantageous and cost effective.

SIX-SACK PEA GRAVEL MIXES

Line pumps cannot handle aggregate larger than pea gravel. Because of pea gravel's small size, professional concrete finishers have concrete batch plants add an extra sack of cement per yard for added strength. Normal 3/4-minus aggregate concrete is mixed with just five sacks of cement per yard. This mixture works great for almost all concrete requirements, although some cities require a six-sack mix for all concrete work done on city property regardless of its planned purpose.

Whenever using a line pump that requires a pea gravel mix or when pouring an exposed aggregate slab with pea gravel, always insist on a six-sack of cement per yard mixture. This extra sack of cement will not only add strength to jobs, it will also help concrete flow through pump hoses and finish off a lot easier. Your concrete fees will be higher because of the added sack of cement per yard, probably about $5 to $7.50 per yard depending on the price of a 90-pound sack of cement.

Don't be dismayed at the additional cost of a six-sack mix and a pumper. Chances are, a professional would have had to hire a pumper too and also pay for the six-sack mix—fees that you would have had to pay anyway on top of his labor fees. Also, almost anything beats wheelbarrowing concrete. Large pours also require a lot of helpers to wheel concrete and a lot of wheelbarrows—six or seven—that you would have to rent. Which is easier and most economical, renting and running wheelbarrows or having the job pumped?

Chapter **14**

Forming curves

*L*andscaping that features a lot of circular planters and rounded flower boxes might be well served by a curved walkway and slab design. In some cases, simply rounding the corners on a patio slab is enough to help concrete flat work blend well with surroundings. For more creative landscapes, free-formed concrete slabs and walkways could flow much better and offer a great deal more in visual appeal. Rounding 90-degree formed corners adds a custom touch to almost any flat work without altering its function (FIG. 14-1).

ROUNDED CORNERS

Forming rounded corners and curves requires special form lumber in addition to 2×4s. Redwood benderboard works well for corners and other rounded designs when it can be supported by 2×4s or plenty of stakes. Many concrete form installers prefer to use 1-×-4 boards for long sweeping curves. You'll also need some ³/₄-inch roofing nails. If benderboard is not slated to completely line the inside face of 2-×-4 forms, you will need a rasp to knock down both ends of the benderboard piece so they are flush with the 2-×-4 face. If not, benderboard ends will cause a notch in the concrete where they protrude out from the 2-×-4 inside faces.

Rounded corners are generally easiest to form after slabs or walkways have already been formed with 2-×-4 lumber. Once all of the regular forms and stakes have been positioned and secured, an 8 or 10-foot section of benderboard is inserted into the corner against the inside face of 2-×-4 forms. With both ends of the benderboard resting against the inside faces of 2-×-4 forms, the benderboard center is pushed into the corner. When you have maneuvered the benderboard's center section into the arc or curve desired, place a stake in front of it to keep it in position (FIG. 14-2).

14-1   A round walkway corner is correctly equipped with seams.

14-2   A section of benderboard has been pushed into a 90-degree formed corner. The stake used to hold this arc will be removed after benderboard has been nailed to 2-×-4 forms and secured.

Secure one end of the benderboard to its 2-×-4 support form with several ³/4-inch roofing nails. Be sure the top benderboard edge matches the top of the 2×4 behind it. Large-headed roofing nails work best to support thin benderboard and their short shafts are not as likely to cause wood to split. Place nails at different locations and heights on the benderboard. Avoid driving nails along the same horizontal line to prevent splitting benderboard.

Roofing nails should start at a far end of the benderboard piece and run all the way down until the board starts to curve and separate from its 2-×-4 support form—about six to eight nails. Be sure the top benderboard edge continues to match the top of its 2-×-4 support form, all the way to the curve.

With one side completely nailed and secure, proceed to the other benderboard end. You can now still move this end toward or away from the corner to change the arc curve. Once you have the arc desired, nail that end of the benderboard to its 2-×-4 support form with 3/4-inch roofing nails. Once both ends of the benderboard are secure, the rounded corner is established. For added form strength and support, fill in the space between the back benderboard arc and the form corner with dirt.

Redwood benderboard becomes very brittle as it dries out. Once dry, it will quickly crack or split as it is forced into a corner. Many professional formers soak benderboard in water before maneuvering it into any curves or rounded corners. For extra tight arcs, soaking benderboard first is an absolute must (FIG. 14-3).

14-3 As benderboard dries out, it becomes brittle and can easily crack or break if forced into a tight arc. Soaking benderboard in water makes wood more pliable.

Once rounded corners are formed, both benderboard ends will be sticking out from the 2-×-4 form faces by 3/8 inch, the actual width of benderboard. To prevent this from causing a notch along the concrete's edge, you could line the entire inside face of all 2×4s with benderboard or simply file down the benderboard ends with a rasp (FIG. 14-4). Use a rasp to file the butt ends of benderboard until they are smoothed at an

14-4 Use a rasp to file down benderboard ends that openly rest against 2-×-4 forms. Cut down benderboard square-cornered ends to a sharp angle so that they flow with the beginning of curves. This prevents notches along concrete side walls and top edges.

angle that allows an edger to cleanly run from the 2-×-4 form to the benderboard. In other words, file down the benderboard lip so it doesn't stick out. Once concrete is in place, an edger will be able to run along both the 2×4 and benderboard forms without obstruction.

MATCHING ROUNDED CORNERS

Some walkways and most patio slabs have more than one corner. Rounding one corner and leaving the other square might look odd. Likewise, having one arc tighter than the other would look awkward. Therefore, all rounded corners on walkways and patios should be as evenly matched as possible to the first arc.

To match rounded corners with one another, you'll need to take accurate measurements. On a rounded corner that has already been formed and secured, measure the distance from the 90-degree corner made by the 2-×-4 forms out to one end of the benderboard. Make a pencil mark on the 2-×-4 form for the new corner to be rounded so that it matches the distance just measured on the completed corner. This will be the spot where one end of the benderboard will be nailed to the 2-×-4 form on the new corner. Take the same measurement for the other benderboard end secured to the completed rounded corner, and mark the other 2-×-4 form for the new corner accordingly. These pencil marks will show where both benderboard ends should be positioned.

Now, on the completed curved corner, measure the distance from the 90-degree, 2-×-4 corner to the center of the benderboard arc (FIG. 14-5). Position a new piece of benderboard into the new corner and

14-5 In order to match rounded corners on slabs or walkways, you need to arc benderboard to identical dimensions.

secure one end so that it matches the pencil mark made for it. Be sure the top benderboard edge matches the top of the 2-×-4 support form. Push the middle of the benderboard into the corner to establish an arc and stake it once it is positioned the same distance away from the 2-×-4 corner as the first corner's arc. Then, check to see if the other end of the benderboard matches up with the pencil mark made for it; it should be close.

Back at the completed curved corner, measure from the 2-×-4 corner to the point where benderboard starts to separate from the 2×4 and make its arc. Mark that same position on the new corner and nail one benderboard side into place. Work with the new arc again to be sure its center is positioned the same distance away from the 2-×-4 corner as the completed one. Move the stake as needed to hold the arc in place. Once that is accomplished, you can nail the other side of the benderboard. Use a rasp to make the benderboard ends flush with the face of their 2-×-4 support forms.

FREE-FORM SLAB

Setting up free-form slabs takes imagination and patience. You can position forms in any design you want with lots of curves and arcs. Tight curves must be formed with benderboard and, because it will not be supported by 2-×-4 forms, you will have to use a lot of stakes. Benderboard's thinness makes it difficult to hold steady in one place and nail it against stakes—patience is the key here.

For long sweeping curves, use 1-×-4 forms (FIG. 14-6). Their thicker width makes them easier to position and nail. They are a little tougher to

14-6 Wide sweeping curves are easiest to form with 1-×-4 lumber. Note the amount of stakes used to keep these in position, both inside and outside the form.

bend but add a great deal more support to your job. You can always make cuts on one face of the 1-×-4 form about halfway through the board to make it easier to bend.

Areas scheduled for free-forming should be leveled out as close to grade as possible, including slope requirements, to make form placement easier. This way, boards won't have to be moved once they are secured in position.

Start your free-form job by placing stakes in the ground about 3 to 4 feet apart and following the basic free-form line you have in mind. Place benderboard along stakes to see how the design shapes up. Use stakes on both sides of benderboard forms to sandwich them in place and keep them supported. Move stakes and benderboard around until you arrive at a free-form design that meets your approval.

A screed board with a level on top of it will work well to help you determine grade and slope. Move forms up as needed to meet grade height or to simply get them out of the way while digging out dirt for grade. Stakes that sandwich benderboard forms should keep boards in place while they are raised. Once grade is established and the bender- board form sits at its correct height, you can use 3/4-inch roofing nails to secure boards to stakes. Nail from the inside face of benderboard forms into stakes positioned along the outside.

To pull these forms when concrete has been poured and finished, you'll have to pull stakes away from forms first until they are separated from the roofing nails, then pull them out at an angle away from forms to avoid the protruding nail shaft. Forms are wriggled loose and pulled horizontally away from concrete. There will be a lot of roofing nails sticking out of benderboard forms, so be cautious.

Securing 1-×-4 curved forms is a bit easier than working with benderboard. Because this lumber is thicker, 6d duplex nails are used the same as if you were using 2-×-4 forms. Be advised, though, even 6d nails are long enough to go completely through stakes and forms to stick out on the inside face of 1-×-4 lumber. So, watch where your foot is braced against forms in order to avoid pounding nails into it (FIG. 14-7).

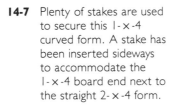

14-7 Plenty of stakes are used to secure this 1-×-4 curved form. A stake has been inserted sideways to accommodate the 1-×-4 board end next to the straight 2-×-4 form.

When free-forming, always remember the basic forming techniques discussed in earlier chapters. Slope and grade concerns are just as valid with free-form work as with rectangular designs. Use your imagination to develop unique schemes but always use your common sense to ensure that slabs are functional and structurally secure.

REINFORCING UNSUPPORTED BENDERBOARD

To help thin benderboard amass more support strength, many form installers double the form's width by sandwiching two pieces together (FIG. 14-8). About the only way to secure these two boards with each other is to put nails completely through them. Nail heads will hold fast on one

14-8 Two benderboard pieces are secured together to make a stronger form for a curved area. Small nails have been driven completely through both boards and are bent over on one side to keep both boards tightly sandwiched.

side and you will have to bend nails over on the opposite side to secure a package together (FIG. 14-9). This double-thick piece of benderboard will be easier to work with and add more strength and support to a curve.

Establishing an evenly flowing free-form arc takes time. Use stakes at each end of the form as braces. These stakes can be moved closer or further apart to create a tighter or wider curve (FIG. 14-10). Once an arc is established, place more stakes to keep benderboard positioned. Don't worry too much about height quite yet; it is more important to secure the correct arc first (FIG. 14-11).

In the case of curved steps, a piece of benderboard positioned in an arc will have stakes on both sides of it for support. The inside stakes are removed after concrete has been placed, at which time the concrete will give it needed support (FIG. 14-12). Once concrete has set up enough to basically support itself, outside stakes are removed and their holes filled with concrete and then finished.

Although wood stakes work great for benderboard jobs, steel stakes can also be used. Nails will have to run all the way through forms and then be bent over (FIG. 14-13). It might take two workers to accomplish nailing with metal stakes. In this case, you might be best off using wood stakes and 3/4-inch roofing nails.

Once stakes are in place and an arc positioned, work to bring forms up to their required height. String is used as a guide to help determine proper form height and slope (FIG. 14-14). Once a side of a form is cor-

14-9 Nail heads hold fast on one side of the benderboard while bent shafts do the same on the other side.

14-10 Stakes located at each end of this benderboard form have been driven into the ground at specific points to indicate where the sides of this step will end.

14-11 Stakes are placed outside this curved form as a means to secure its shape and support it for the pour. Form height is adjusted after all stakes have been securely pounded into the ground.

14-12 Stakes placed inside formed areas are pulled after concrete is poured. This stake was used on the inside of the arc to keep a benderboard form secured against a small, 2-×-4 block.

14-13 Nails easily penetrate thin benderboard. A hammer is placed against the nail head for support while the nail shaft is bent over to secure it against the stake.

14-14 A 4-foot carpenter's level is used to determine this curved form's height. Note the string used as a grade guide.

rectly adjusted, use a level to be sure the other side matches. When that is done, check the entire form length to ensure there are no weak spots. Use as many stakes as necessary to guarantee wet concrete will not push boards out of position.

With step forms correctly positioned and secured, you must check grade level to make sure it is not too deep. Center step portions need not be deeper than 4 inches. On the sides, however, you should leave an open area that reaches down to the rest of the walkway or slab surrounding it. This open space allows for a solid concrete structure to encompass not only the main body of the job, but it also adds a great deal of support to steps. The extra concrete used around step perimeters works well to prevent erosion from occurring under them.

CUSTOM BORDERS

Real custom work shines through when form installers design curved borders inside rectangular slabs. Once concrete has been placed and finished, jobs will appear as if a curved slab is outlined with a rectangular border. These types of custom slabs generally feature borders one color and slabs another or one an exposed aggregate and the other etched. To produce this type of custom work, however, you must have a definitive work plan in mind and be on your toes throughout the entire pour (FIG. 14-15).

The first task at hand for a job like this is to form up the rectangular slab, or the outside forms if they are to be curved with a rectangular border inside. Once forms are positioned and secured, the grade must be properly leveled and prepared. Only after that can work begin with benderboard to create an inside border.

14-15 This custom concrete job features a curved front porch step landing and wide entry slab, both outlined with borders.

Establish curved ends first using stakes for support. Once the right arc is achieved, secure benderboard to stakes using 3/4-inch roofing nails. Use a screed board to determine proper form height. NOTE: for inside borders, redwood benderboard will permanently remain inside concrete—therefore, redwood stakes are driven down below the top of benderboard forms to allow plenty of concrete to cover them (FIG. 14-16). These stakes must be at least 1 to 1 1/2 inches below form tops. If the ground is too hard to achieve that, use a saw to cut stakes to a correct length.

14-16 A screed board spans this formed area to help determine the benderboard outline form's height position.

With the curved portion complete, you will need to run a straight border along the side forms. Use benderboard to match that which is already in place for the curve. Placing benderboard in a straight line is easily accomplished with the help of blocks (FIG. 14-17). While placing stakes, brace your foot against the stake to sandwich the form between it and the block. The block will be supported on its other end by the 2-×-4 form (FIG. 14-18). These blocks must be left in place until concrete is poured. Once concrete surrounds blocks, they can be gently pulled while forcing concrete into the open space created by their removal. This will ensure that benderboard is not moved out of position, as concrete surrounding it will hold it in place.

14-17   Short, 2-×-4 blocks support a benderboard outline form in a straight line in preparation for staking.

14-18   Short redwood stakes are pounded into the ground in front of each block.

BRACING BENDERBOARD

Benderboard can be used for a multitude of purposes when forming custom concrete flat work that features curves, arcs, and other unique designs. Its pliable nature lends itself to creative shapes. But, because it is so thin, it will need plenty of support to hold back the weight of wet concrete. You already know about soaking dry benderboard, but what about reinforcing it in preparing for a concrete pour?

Basically, there are three ways to reinforce and support benderboard concrete forms: lots of stakes, 2-×-4 form supports, and dirt. Placing wood stakes every 1¹/2 feet along benderboard forms is a good idea. This is especially true when benderboard is actually used as a form and will not be supported on its backside by a 2-×-4 form or concrete.

If long pieces of benderboard are used as a border or an expansion joint, you'll need to use a lot of stakes to keep wood straight. Some professional concrete finishers prefer to drive redwood stakes into the ground about 1¹/2 to 2 inches below benderboard tops, nail the benderboard to them, and just leave the whole package buried in concrete (FIG. 14-19). Others might place stakes directly across from each other to sandwich benderboard between them. This method works well but still requires stakes about every 1 to 1¹/2 feet to keep benderboard straight. Even with extra stake support, concrete must be poured slowly and gently against benderboard on both sides at the same time to ensure it maintains its straight position.

Another way to brace benderboard is with short, 2-×-4 boards. Even free-form jobs have short straight sections that can be reinforced with a short 2×4 and a couple of stakes (FIG. 14-20). Brace with a 2×4 and fill in

14-19 A long benderboard expansion joint is being secured at the end of a new driveway where it meets with a sidewalk area. Redwood stakes will be left in place under concrete.

14-20 Short, 2-×-4 boards will be used to brace benderboard. Dirt can be placed between the benderboard and 2×4s for added support.

14-21 An inexpensive and easy way to support benderboard forms is with dirt or sand. This corner is adequately reinforced with dirt so that the curve will maintain its shape during the pour.

voids with dirt along shallow arcs. Be sure to have plenty of 2-×-4 material and stakes on hand for this type of job.

The least expensive, quickest, and easiest way to brace most benderboard forms is with dirt (FIG. 14-21). Damp dirt or sand provides ideal support. Dirt and/or sand cannot take the place of stakes, but they can add a great deal of support to forms that must hold back the weight of heavy concrete. In most cases, dirt or sand packed to just below the top of benderboard forms and back a good 6 to 8 inches is sufficient.

Chapter **15**

Stringer designs and installation

*T*he term *stringer* simply refers to a board encased inside a concrete slab or walkway. Stringers do a lot to break up solid concrete by dividing slabs into eye-appealing custom sections (FIG. 15-1). Not only are stringers used as simple decorations, they double as useful expansion joints.

Because wood stringers are encased in concrete, they are nearly impossible to replace should they rot or otherwise fall apart. For that reason, only pressure-treated lumber or redwood is used. If neither types of wood are available in your area, consult a lumberyard salesperson to find out which type of wood is available that will resist mold, mildew, and dry rot while in place as a stringer.

DESIGN

Stringer designs are almost limitless. Just about any pattern imaginable can be produced (FIG. 15-2). Any custom concrete finish desired can be applied next to stringers. What makes stringers even more appealing to professional concrete finishers is that they can be used as supports for screed board operations. This feature alone can save the work necessary to install screed forms or the need for wet screeding.

You can keep stringer designs simple by just placing one pressure-treated 2×4 in the center of your slab to divide it into two equal halves. For a greater design, you could use three or four stringers to section your slab into several separate sections. You can even incorporate enough stringers to divide a slab into 30 or 40 equally sized or dimensionally varied squares or rectangles and leave a few of them open for decorative planters. Further, stringers do not necessarily have to run perpendicular

15-1 Stringers have been used to outline open planter and mulch areas.

15-2 Rock salt has been pounded onto a fresh concrete slab, which incorporates a simple stringer design.

or parallel with each other. They can be cut and installed to form diamonds, octagons, triangles, and so on.

You should have a realistic idea of what it is you are trying to design before actually nailing forms and stakes together. Spend time with pencil and paper sketching out various stringer patterns and how they will be assembled. Certain designs require boards be notched and fitted into one another. Without a clear plan, you could end up notching boards in the wrong place. Pressure-treated lumber and redwood are quite expensive, so be sure each cut is done correctly.

To get a good perspective of what your penciled plans will look like, loosely set out stringers in the designed pattern before doing any cutting or nailing. This way, should you decide to change a few board positions, you won't have to pull nails and you won't have made any miscalculated cuts.

FIRST STRINGER

The first stringer placed should be the one that must have the most support and strength. This could be one that splits a slab's length in half and is nailed at both ends to side forms. If it is to be centered on side forms, simply measure them to find their exact center; be sure to measure from the same reference point, such as from the house out or from the end form in. This will ensure that your pencil marks on top of both side forms are equally positioned in the middle of the slab area.

Lay a stringer next to the pencil marks on side forms. Check the stringer to make sure its top edge is not sticking up higher than side forms. If it is, the grade is too high and some dirt will have to be removed. Also, be sure the stringer is not too long for the space it is to occupy. Many boards are not cut exactly to size and might be an inch or two longer than expected. If that's the case, simply measure the distance to be covered by the stringer and cut it to fit.

Once the stringer is in position and the grade correct, nail one end of the stringer to a side form. Be certain the center of the stringer is in line with the side form's center mark. Drive nails from the outside face of the side form into the end of the stringer. Be sure duplex nails are long enough for the job. After the first stringer side is secured, confirm its proper grade height and position and then nail the other side.

Even though both ends of this stringer are secured to side forms with two nails in each end, you need to stretch a string from one end to the other to use as a straight-edge guide. Push the stringer out or pull it in as needed to align it with the string and drive stakes in the ground to keep that stringer straight. These stakes should be placed about every 4 feet or as often as necessary to keep the stringer in line. You should not need to nail them but you could sandwich boards between stakes to keep everything straight.

On shorter stringer runs of 12 feet or less, you might only need to place stakes on one side of a stringer. Put them on the side that will receive concrete last so they will support stringers while concrete is poured against their other side. Longer stringer runs must be more heavily supported with stakes on each side.

Extra long stringers that need a lot of support and that will have wheelbarrows wheeled over them might be best supported with stakes and nails. In this case, drive redwood stakes into the ground on both sides of a stringer until they are about 1 to 1$^{1}/_{2}$ inches below the stringer's top; then secure them to the stringer. This allows concrete to completely surround stakes and eliminates the need to pull them after the pour. If additional stringers are to be placed over this one, position stakes so that they will not be in the way.

Should your plans call for another stringer to intersect this lengthwise one at its center, use a tape measure and pencil to locate and mark the exact center of the secured stringer. On each side of the center mark, measure out $^{3}/_{4}$ inch and use your square to make a straight line on both $^{3}/_{4}$-inch marks. The distance between these $^{3}/_{4}$-inch marks should be exactly 1$^{1}/_{2}$ inches—the width of a 2-×-4 board. This designates the location where the intersecting stringer will go.

Anchors

Before setting the second stringer, place anchor nails along both sides of the secured stringer. Partially drive 8d or larger nails into both sides of the stringer at 2- to 3-foot intervals. Allow at least half of each nail's length to stick out. These nails will be encased in concrete and will hold the stringer in position after concrete has set up.

Although it might not happen often, stringers without anchors have been known to float up out of slabs after a year or two. Raised off a slab's surface just a little bit is plenty to cause a definite tripping hazard to those walking near it. Anchor nails prevent stringers from floating up and should be put into both sides of all stringers.

SECOND STRINGER

As you did with the first stringer, measure the second stringer to ensure it is cut to proper length. This second stringer is supposed to separate the slab's width and will intersect with the first stringer at its center point. Lay the board down first to see how it fits. It will ride high on top of the first stringer because neither of them have been notched yet. If all's well, find the center point on the second stringer and mark it. Use your square to make a straight line at the $^{3}/_{4}$-inch mark on both sides of the center mark, just as you did to the first stringer.

Place the second stringer on top of the first and line up the $^{3}/_{4}$-inch marks located just to the sides of the center. If they all match, you are about ready to make notches (FIG. 15-3). Set up your square to measure down the face of each stringer 1$^{3}/_{4}$ inch from all four $^{3}/_{4}$-inch marks. This will put you at the center of each face. The 1$^{1}/_{2}$ inch on the top will be for the stringer thickness and the 1$^{3}/_{4}$ inch on the stringer face will be for its width. Cut along both lines made from the top of the stringer to the center of its face on both sides of the center.

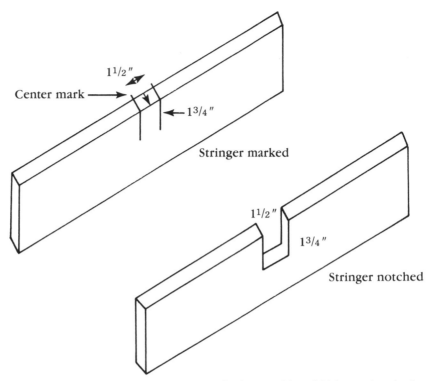

15-3 Intersecting stringers should be notched to provide solid joints and maintain board straightness.

After both cuts have been made halfway down the stringer face, lay the board down. Use a hammer to firmly smack the chunk of wood located between the two cuts. This should quickly knock the chunk of wood out of the stringer. Use a chisel to clean the notch.

On the stringer already placed and secured, make the same cuts and knock out the chunk of wood thereafter. Use a chisel to clean out burrs from the notch. You should have two notches identical in size to each other—1½-inch wide and 1¾-inch deep—when finished. These dimensions should allow stringers to intersect each other and lay flat.

Pick up the second stringer and place its notch over the first stringer's notch. Push down on the free stringer to force it into place until its top is flush with the first stringer's notch. You might have to gently tap it with a hammer, as this should be a tight fit. If it is too tight, use a rasp to widen or deepen either notch. Secure notched boards with a nail through the middle. Nail from the side, not from the top, as a nail from the top will be visible and unsightly.

With the second stringer firmly positioned in the notch of the first, find the center point of the end form and nail the stringer to it with two big duplex nails (FIG. 15-4). Staking and anchoring maneuvers are the same as for the first stringer (FIG. 15-5).

15-4 Stakes are used to support stringers during concrete pours. They can also be used to take out slight bows in stringer lumber.

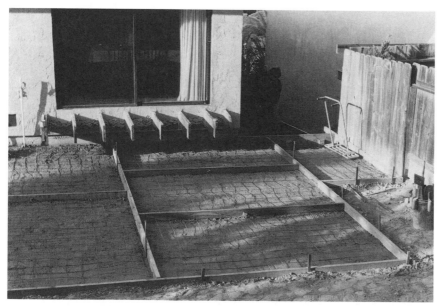

15-5 Install the longest stringers first. They are generally the ones that can be easily nailed to outer, 2-×-4 forms. Once they are secured, install shorter stringers. The stringer running from the house to the end form on this job was installed first with the shorter ones installed afterward.

Although notching might take more time and involve intricate work, it makes a strong joint and is much more precise than just butting the ends of two separate, but shorter, stringers to the first. Notching guarantees that joints will be square and side runners will not get moved during a pour. It is also easier to notch intersecting boards than to accurately measure, cut, and stake two boards in a straight line.

All fully intersecting stringers should be notched. Whenever possible, secure the first board so it makes a solid run and is nailed at each end for support. This way, you'll have a rigid and secure point to work from and can gauge accurate measurements and notching locations from it.

UNIQUE STRINGER PATTERNS

When purchasing 2-×-4 stringer lumber, make sure at least one of the 1 1/2-inch sides is perfect on each board. Knots, cracks, gouges, and other imperfections that will show through on top of stringers are unsightly. These types of problems will not be visible on the 3 1/2-inch faces and the bottom 1 1/2-inch side because they'll be buried in concrete. But that one 1 1/2-inch side will show and you will want it to be perfect. Along with that, carefully examine each board to make sure it is straight. Crooked, bowed, and twisted boards are very difficult to work with when trying to form sets of straight stringers.

If yours is a truly unique design, try to buy stringers at a size that will be most accommodating and easy to work with. Obviously, if your slab is going to be 36 feet long, you won't be able to easily find or transport a 36-foot 2 × 4, but two 18-foot or six 6-foot boards might work out just right.

To demonstrate the basics in putting special stringer designs together, let's use a 12-×-36-foot slab that will be divided into six separate sections (FIG. 15-6). The two stringers that run from the house to the end form will

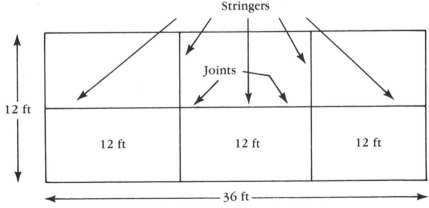

15-6 Twelve-foot stringers are best suited for this design. The two stringers running from top to bottom would be installed first because their ends can be securely nailed to an end form. The three stringers running from side to side will fully span distances between intersecting stringers so that no free-floating joints are created. All stringer ends will butt against another stringer.

be 12 feet long. If you used just two 18-foot stringers for the 36-foot span, they would have to butt together openly in the middle of the slab. Should you use three 12-foot stringers instead, each of their ends would butt next to or between stringers or end forms perfectly. With these, you would not end up with a floating joint in the middle of the slab and you would have a much stronger stringer framework.

This basic board-size logic applies to all form and stringer purchases. Determine beforehand what lengths will be needed. There is no valid reason to purchase more lumber than required for jobs, while money spent on unnecessary lumber takes away from the money you are saving by doing the job yourself.

When designing stringer patterns, consider leaving some open sections of various sizes for use as planters (FIG. 15-7). If open sections are planned for, don't forget to deduct their square feet calculations from your overall concrete yardage figures. Simply measuring the length and width of a slab with open planters without deducting for open space will cause you to order more concrete than needed. You'll have to pay for concrete ordered but not used.

15-7 This custom patio slab incorporates stringers, open planters, and a short curved section. Notice how stringers are installed to equally span the width of a bar installation to the right.

PLACING CONCRETE INSIDE STRINGER DESIGNS

The most significant problem associated with pouring concrete inside stringer designs is that stringers get in your way. More than one finisher has tripped over stringers. Take your time pouring these types of jobs to prevent injuries and unnecessary stringer damage.

If your job is going to be wheelbarrowed, you have to build ramps over stringers and place extra stakes next to those stringers that will be wheeled over. Try to figure out the most logical routes to take so that the minimum amount of stringers are affected. On large pours over 5 yards, consider a concrete pump. This will eliminate any wheelbarrowing over stringer maneuvers.

Build ramps out of 2×4 scraps and plywood. Try to make them in such a way that the wheelbarrow tire does not touch stringer tops. You must also put stakes on both sides of stringers at each end of the ramp. Remember, wheelbarrows will be heavy and will hit ramps with a great deal of force. Anything you can do to rigidly support stringers will greatly help them to stay in place, avoid damage, and remain straight.

When pouring concrete into stringer designs directly from concrete truck chutes, be very aware of every step you take and how the chute is worked back and forth. Stakes might stick up high enough to interfere with chute movement, so watch your step.

Use screed boards directly on top of stringers. If certain sections are formed smaller than others, use a shorter screed board. Make the job as easy on yourself as possible. And don't forget which direction you plan concrete to be poured from, especially if stakes were only placed on one side of stringers to support them from only one direction.

EDGING AND CLEANING STRINGERS

To make jobs look clean, crisp, and professional, both sides of stringer tops must be edged. Edge them just as you would perimeter forms. This dresses up the job with clean edges and gives you a very good perspective of how concrete meets up with stringer tops. Perhaps low spots need filling in and high spots reduced.

For long slab stringers, have a walking edger with extension poles ready at hand (FIG. 15-8). This tool will make your work much easier. The only way to edge long stringers with a hand edger is to tiptoe on stringers and edge at the same time, a very difficult chore. It also results in footprints along stringers and other related problems.

Jobs with more than 3 or 4 stringers should have at least one helper designated for edges only; after the pour, that will be his only responsibility. Edging is an important and time-consuming task that should not be taken lightly. Once edges start to get away from you, a lot of hard, vigorous, and speedy work must be done to get them back in shape. Let one helper concentrate on edges while others work the tamp, bull float, fresno, and the like.

Stringer jobs with 30 or 40 small squares or rectangles will require the total concentration of at least two helpers who do nothing but edger work. By the time they complete an initial edging chore, concrete will probably have set up enough to require a second edging. By the time that's done, they'll have to start over again.

By all means, attempt to clean stringer tops as you edge them. This helps to reduce the amount of concrete debris that finds its way to slab

15-8 All stringer edges must be edged. If not, concrete surfaces look unfinished and unprofessional.

surfaces during bull float and fresno work. Use a putty knife to scrape as much as can be reached from outside forms. For long stringers, you might be able to maneuver the side of an edger in such a way as to scrape off some excess concrete aggregate without messing up edges. Once a slab has set up and you are finishing with knee boards and hand trowels, make a concerted effort to clean stringer tops to prevent concrete residue from hardening on them.

PROTECTING STRINGER TOPS

There is no reasonable way to avoid concrete getting on stringer tops. Some finishers have used heavy-duty duct tape on them for protection but were dismayed when tape broke loose and wandered into concrete. Others prefer to paint on a protective sealer before a job is poured to keep concrete from firmly attaching itself to wood surfaces (FIG. 15-9).

Along with tape coming loose from water absorption, operating screed boards on top of taped surfaces causes the material to rip, tear, and fall off. When loose tape is not noticed early, it can really cause finishing problems while using a fresno. Later, extra work is needed to remove embedded tape and repair the damage left behind with the use of hand trowels.

Cleaning stringer tops while doing other related chores works well. While tamping, stop for a moment and use a hand float to scrape off excess concrete. Either toss that excess aside or tamp it into the slab. Carry a putty knife with you while out on the slab with knee boards and take time to clean stringer tops after perfecting their edges. Once concrete has fully cured, in about two weeks, use a wire brush to remove whatever concrete residue is left.

15-9 Stringer tops are the only parts of the board that will be visible after concrete is poured. Consider coating them with a sealer designed to resist concrete adhesion.

15-10 Some concrete finishers use motor oil to protect stringer tops from concrete. Wood stains and other products specifically designed for this purpose may be better to use.

Special products used to seal stringer tops are available at lumber yards, concrete plants, and some hardware stores. A few finishers have even had luck using motor oil and diesel fuel to prevent concrete buildup on stringer tops. This may be fine, but you have to wonder what stringers will look like afterward and whether or not they will accept paint or stain after being coated with petroleum-based materials.

If you have decided to use clear redwood for your stringers, don't take a chance on using an inferior product to prevent concrete build-up and protect stringer tops. Instead, check with your local concrete dispatcher to see what specific products are recommended for your regional location and climatic conditions (FIG. 15-10).

Stringers that are securely braced and supported make wonderful screed forms to make screeding jobs a breeze. You might incorporate a stringer into your plans just for that reason. Along with making screed work a little easier, stringers make good expansion joints and also add a touch of flair to otherwise plain concrete flat work.

Chapter **16**

Steps and stairways

*F*orming, pouring, and finishing steps and stair-
ways requires ingenuity and attention to detail. You must have a very
good idea of how an entire job is supposed to go together before you start
sawing boards, driving stakes, or pounding nails. Small stoops or rectangular steps with only one or two risers are generally simple and straightforward to form and finish. Custom designs and large scale stairways,
however, can often be troublesome. Setting the correct slope, step height
and width, and actual form installation techniques are difficult to master.

Individual steps seldom rise more than 7 inches. In fact, most of us
are quite familiar with 7-inch steps and are caught off guard when we
have to step up higher than that; it feels awkward. Likewise, horizontal
step runs (width), should never be less than 10 inches, with 12 inches a
much more practical minimum. Runs of less than 10 inches do not allow
enough room for people's feet to fit securely. Depending on the slope
and distance to be covered, step runs can certainly be as wide as desired
as long as they are all the same or at least evenly split into symmetrical
sections (FIG. 16-1).

Step height can always be less than 7 inches but should not be so
short that it becomes somewhat invisible. A 2-inch step, for example, is
just too short for people to clearly recognize and will probably cause
most folks to trip. Therefore, plan on making steps rise anywhere from 4
to 7 inches. This makes them tall enough for everyone to clearly recognize and yet not so tall as to feel uncomfortable or cumbersome when
being walked on (FIG. 16-2).

Novice form installers and concrete finishers are encouraged to
watch a professional put together a set of steps before tackling similar
projects on their own. A concrete dispatcher should be able to locate a
contractor scheduled to pour steps or stairways.

Custom designs and large pours must be formed securely or concrete
will cause forms to "blow-out," or break loose. When forms break on

16-1 Step height and run must be practical. Although these steps are offset, their walking surface is functional and comfortably spaced.

16-2 Step height should be no more than 7 inches. These forms support concrete designed to hold back a bank located next to a patio slab and will not be used as steps.

these types of jobs, concrete has to be shoveled out and new lumber installed as fast as possible because previously poured concrete will have already begun setting up. Along with the frustration of dealing with such dilemmas, standby time for concrete trucks could become significant.

INITIAL LAYOUT

Before starting your step or stairway project, examine other jobs to see how they are set up. Note the location of top steps and how far they extend horizontally before dropping off to the next step. How is the grade held back on the sides of steps? Are the steps poured on top of the grade or are they sort of sunk into it? Does the bottom step drop off directly onto a walkway or patio slab or is it actually poured at the same level as the concrete it butts up to?

Ideally, top steps should extend out horizontally as far as possible while still being flush with the pathways they are extending from, much like a landing or an extension of the surface from which they come (FIG. 16-3). In other words, the first step should not immediately drop off directly from a pathway or other walking surface. People should be able to step flatly on the top part of the stairway first before having to drop down. The stairs should drop down in succession after that.

16-3   The top part of this stairway extends out horizontally as far as possible to create somewhat of a landing before dropping down. Notice how its front side is formed at 4 inches deep, just right for a future walkway addition.

At the bottom, the last step should also be somewhat of a landing. It should go in toward the hill at the same level as the walkway or slab it butts next to. It should extend in at least as far as the width of all the other steps so that it remains equal in dimension. This prevents people from having to drop down from the very last step right onto the adjoining walkway or path surface. They will have a step-wide landing to step on before actually setting foot on the bottom level at the base of a hill. Also, the bottom step will not abruptly rise off the edge of the lower-level path because its rise will start about 12 inches inward.

Once top and bottom steps (landings) are formed in, grade is set for the rest of the steps. Keep in mind that the step-face form for the top landing will be the starting point for the stairway drop and the bottom edge of the step-face form for the last step will be at the level of the landing below. The distance between these two points is what will be measured and equally divided into actual step rises and widths.

For example, let's say that the vertical drop is 28 inches, which is determined by holding a board with one end resting on the top surface while the free end extends out toward where the steps will stop. With the board level, measure from its bottom side to the grade below. This will show how far the steps must drop. For a 28-inch drop, you would set up forms to accommodate four, 7-inch high steps.

Now, to determine the step width, simply measure the horizontal distance from the top landing's step-face form to the bottom landing's step-face form. You will have to hold your tape measure in the air and measure straight out horizontally. For this example, let's say the horizontal run is 4 feet.

Now, with a 28-inch drop and a 4-foot run, you could easily calculate four equal steps 7 inches high and 1 foot wide. This is done by dividing 28 inches by 7 inches (the maximum step rise), which equals 4; this job requires four, 7-inch-high steps. The horizontal run of 4 feet is divided by the necessary 1 foot width to equal four 1-foot-wide steps. This very basic formula should work well for most small step jobs.

FORMING

Side forms must be wider than step heights. This is because concrete must flow under each step so that the entire structure is tied together. If your steps are going to rise 7 inches, use 2-×-10 lumber to form sides. With the step-face form tops positioned to match side form tops, there should be a 2- to 2 1/2-inch gap below step-face forms where concrete can flow together as one unit. The step-face forms will have to be cut to exactly 7 inches.

Dig trenches from the top landing form to the bottom landing form where you plan to insert side forms. These forms will be secured before actual grading for the entire job is accomplished. Position side forms so that their top ends butt up with top landing forms. NOTE: you will have to cut the ends of the side forms at an angle so they will fit flush; the tops of the side forms should match the tops of the top landing forms. The bottom ends of the side forms will rest on top of bottom landing forms (FIG. 16-4).

Once forms are in position, secure them with enough stakes to keep them steady. Do not securely stake them yet. First, mark locations for

16-4 This stairway has been formed and will now be reinforced with plenty of kicker stakes. Note that the top side form end was cut at an angle to match the end of the landing form and that its other end simply rests on top of the slab's side form at the bottom.

step-face forms so that succeeding stakes will not be placed in the way of nailing step-face forms to side forms.

At this point, put in the top, step-face form at the end of the top landing (FIG. 16-5). Remove dirt as necessary to get the form to fit correctly. Nail through side forms into step-face form ends with large duplex nails.

16-5 The top landing was formed first followed by a side form running downhill. The form at the front end of the landing will actually be the top step-face form, as this is the first step to drop down. This form will be pulled early so that concrete used to make the step-face can be finished.

Be sure to use duplex nails because these forms will have to be pulled early in order to finish concrete step-faces. With the top step-face form in place, use a square-point shovel to notch out dirt just in front of, behind, and under, the form. This is so concrete can be about 3 to 4 inches thick as it makes its way down to the next step.

Continue to dig until the second step area is level. Measure from the step-face form the distance needed for the width of the next step. In this example, that would be 1 foot (FIG. 16-6). Mark as needed and then place the next step-face form. The grade itself should look like a set of steps running just inside the step-face forms to allow room for concrete (FIG. 16-7). Before nailing step-face forms, use a level to make sure the top side of the lower form is just a shade lower than the bottom of the step-face form above it. Place the bottom edge of the carpenter's level in line with the bottom edge of the higher step-face form. Rest the other end of the level on top of the next lower step-face form (FIG. 16-8). Continue this process until complete.

Once all of the forms are in position, use plenty of stakes to secure side forms. Kicker stakes are needed for all forms taller than 6 inches. This

16-6 Successive step-face forms are installed from top to bottom.

16-7 Grade under step forms has been accomplished in somewhat of a step pattern.

helps them to remain upright throughout the pour. Next, place stakes in front of step-face forms to keep them from bulging out when concrete is placed behind them. Once concrete begins to set up, these stakes will be removed and their holes filled with some concrete that you have set aside for that express purpose.

Now, if you want to save yourself some concrete finishing work, cut the bottom edge of step-face forms at a 45-degree angle. Do this before placing forms. Have the open-angled section point out. This allows plenty of room for you to operate a hand trowel over the entire width of the steps without having to reach under the form (FIG. 16-9). Another tip is to coat the inside face of step-face forms with motor oil to help prevent concrete from sticking to the wood as forms are pulled away.

POURING CONCRETE

Concrete yardage for steps is more than that which is used for normal, 4-inch slabs. This is because there is open spaces under step-face forms and an inside trench along side forms to prevent erosion under them. Carefully measure and calculate these areas for concrete yardage and then

16-8 The bottom of this carpenter's level is positioned in line with the bottom of the step-face form above and on top of the step-form below.

16-9 The bottom edge of this step-form was cut at a 45-degree angle to allow an open space for hand floats and trowels.

16-10 Chunks of the old porch stoop have been put inside this newly formed step area to take up space so that concrete will not have to be much more than 4 inches deep.

plan to order at least ¼ yard more to allow for any mistakes (FIG. 16-10). This is common practice among concrete finishers. They always prefer to have more concrete on hand than not enough. Confirm your dimensional measurements and concrete yardage estimates with the concrete dispatcher.

Concrete should be a little on the dry side when pouring steps to prevent excessive downhill slump. Wet concrete will quickly flow toward the bottom to bulge lower steps while top steps have trouble maintaining a flat, level surface. Many finishers like to pour a couple of inches of concrete over the entire base before starting to fill areas up to the tops of forms. This allows a shallow base of concrete to set up a little in hopes it will, in turn, serve as a base for the rest of the load.

Most finishers prefer to fill in forms at the bottom first and then move upward to the top. Again, this is to solidify lower levels first so that the upper ones can be supported and not slump downward. A square-point shovel is used to help push concrete into lower open spaces for complete coverage. Pouring concrete steps and stairways must be done rather slowly and under complete control at all times.

An enormous amount of concrete weight will be pushed against step-face forms, and pouring too much too fast can cause forms to blow out. Fill in forms about an inch at a time and slowly build up toward their tops. This will ease the amount of pressure against forms as levels of concrete gradually increase.

Because step faces must be finished just like flat work surfaces, try to ensure that there is enough concrete cream located next to forms. You can do this by sliding a thin piece of benderboard up and down along the inner side of step-face forms to push aggregate away and allow cream to fill in. Do this as you are pouring the job. You should also tap form faces with a hammer to vibrate aggregate back and cream forward. A quick series of continued light taps works much better than a few solid blows. Most finishers do both of these things to ensure plenty of cream is available when step-face forms are pulled and they begin finishing with hand trowels.

Steps are always cautiously tamped. Vigorous tamping with a lot of force adds even more pressure to already-stressed forms. Tamp just hard enough to push down aggregate and bring up cream. It is much better to tamp lightly a few times than once with a lot of force.

Bull float, fresno, and edging operations are carried out just as with other flat work. You might have to maneuver the bull float and fresno without extension poles for best maneuverability. Simply place them flat on steps and draw them out toward the downhill side. Use hand floats and trowels as necessary.

FINISHING STEPS

Although finishing concrete on steps is completed the same as for any other slab, there are a few unusual chores unique to steps. For one, stakes embedded in concrete as supports for step-face forms will have to be removed after concrete has sufficiently set up enough to support itself inside forms. Holes left behind have to be filled and spots finished off to match the rest of the surface. In addition, step-face forms will have to be removed while concrete is still somewhat workable in order to apply a decent finish to them.

If step-face forms are pulled too late, concrete might be too hard to finish as desired, leaving behind an unattractive permanent impression (FIG. 16-11). This condition can be eliminated by cutting the bottom edges of forms at a 45-degree angle to allow room for trowel work. Pulling forms too late can result in step faces setting up too hard to accept a decent finish.

The timing of pulling step-face forms is critical. Pulled too early, concrete will slump down and possibly fall away from steps; pulled too late means a lot of vigorous finishing work. The best time to pull stakes is usually after the last fresno application. Concrete will still be too wet to hand finish but should be strong enough to support itself against step-face forms without the need of stake support. It should also accept the efforts of filling in stake holes with no problem.

Remove nails and gently loosen stakes with light hammer taps. Hitting them too hard will damage surrounding concrete. Pull straight up on them with the least amount of sideways movement as possible (FIG. 16-12). You should have some creamy concrete readily at hand that was purposely set aside in a bucket to be used for filling stake holes. Use a small masonry trowel or putty knife to fill holes (FIG. 16-13).

When holes are filled with concrete, use a trowel to smooth the surface (FIG. 16-14). Do an adequate job without spending too much time in the process—the entire step will have to be hand troweled later anyway. Concentrate on pulling stakes and filling holes, using knee boards as necessary (FIG. 16-15). Pull all stakes from all the step-face forms before pulling

16-11   Step-face forms were not cut at a 45-degree angle for these steps, as evidenced by the 1¹/₂-inch-wide blemish running along the base of the step.

16-12   Pull stakes straight up when removing them from step forms after concrete has initially set up to prevent damage to surrounding concrete.

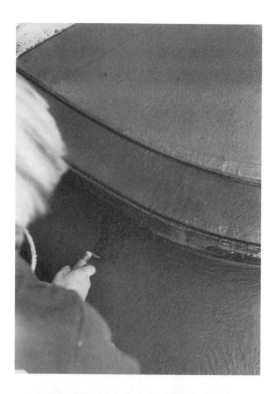

16-13 Save some cream to fill in empty spaces created by removing step-face form stakes.

16-14 A small hand trowel is used to fill a stake hole with creamy concrete.

16-15 A curved benderboard step-face form has been removed from this custom concrete front porch step. Concrete is hard enough to allow knee board use without suffering intense dent impressions.

off any forms. This way, forms can continue to hold in moisture until you are ready to finish the concrete behind them.

Concrete resting next to forms will hold more moisture than uncovered surfaces open to the air. Therefore, once step surfaces are about ready for final finishing, the step faces ought to be set up enough to support themselves without forms. Starting at the top, pull nails out from side forms that secure the ends of step-face forms, then gently tap step-face forms with a hammer to break them loose. Go slow and gently. Pockets of concrete sometimes stick to forms and are pulled away as forms are removed. This problem is lessened if inside faces are coated with motor oil before the pour.

Slowly pry forms away from concrete pulling straight out horizontally and using light hammer taps to help break their seal. Do not pull up on forms. You could pull chunks of concrete off the edge. Once forms break free, use a metal trowel to scrape off any cream residue, then spread it on faces as you finish them. If the edge is not perfect, hold a short 2×4 against concrete and use an edger as needed. Finish step faces as you would any other concrete.

If open pockets, or holes, are located on step faces, simply gather some creamy concrete from the tops of forms or elsewhere and spread it on, just like you would patch a hole in a wall. Once the face is finished smooth, run a foxtail broom across it to effect a broom finish (FIG. 16-16). Once the face is finished, complete the rest of the step. Then move on down to the next one and do the same thing.

If you find that concrete is still too wet to finish, smooth blemishes and then leave it alone. Do not overwork wet concrete step faces. They are fragile without form support and too much trowelling could cause them to slump down or break away. Should that happen, don't be afraid to put forms back and leave them there until concrete sets up better. On

16-16 This step face is being finished with a foxtail broom.

the other hand, if you find faces hard to finish, dip your trowel in water and apply it wet to help moisten mud. If necessary, wet down the foxtail broom to effect the final finish.

Caution: if your steps are scheduled for an exposed aggregate finish, remove stakes at their prescribed time but do not remove step-face forms until all of the other surfaces have been completely exposed. The reason for this is simple. The amount of water that will be used to expose the aggregate on flat surfaces will just erode edges and step faces. Forms help to prevent erosion and keep faces properly supported. If possible, delay exposing step faces until the last minute. You might have to work harder with a broom but you'll lessen the risk of ruining otherwise decent step faces. Go slow and use the least amount of water as possible.

Chapter **17**

Other considerations before pouring

A lot of novice concrete finishers have realized too late that once concrete is poured and finished, alterations of almost any kind are very difficult to accomplish. Such would be drainage piping under slabs, electrical and water conduits for accessories near slabs, custom-forming designs, and so on. Ideally, you should have long-range landscaping plans in mind when designing and forming your concrete work. This way, you'll be able to install drainpipes as needed and any conduit required to accommodate future lawn sprinkler systems, yard lighting, or perhaps a natural gas line to serve a future barbecue grill.

Discuss landscaping and overall home improvement plans with family members. Their comments and ideas might spark interest in areas you never considered. Once you have a set of long-range landscaping goals, form and pour your concrete slab or walkway to include all the attributes needed to accommodate future projects.

SOME FINAL FORMING HINTS

Steps and walkways can be custom formed and still be practical (FIG. 17-1). Some offset and staggered step patterns break up solid concrete yet are still functional. Leaving pressure-treated forms in place and staining them later to match house trim or adjacent deck colors might also be creative and practical at the same time. Determine the actual use for concrete additions and then imagine how you can vary that design to better blend with surrounding landscape shapes, grade slopes, and future improvement projects.

Carefully inspect your concrete job site the day before pouring. Tap forms with your foot to be sure they are secure and positioned correctly.

17-1   Custom formed steps lead to a concrete deck surrounding an in-ground hot tub.

Check for bowed forms on long runs. Place a stake against any bowed section's inside face to straighten it out (FIG. 17-2). Remove stakes after concrete has been poured around them, as concrete weight should be
enough to hold back the bow. This is a common practice, especially for
curved benderboard forms.

Butting thin form faces flush with 2-×-4 material is sometimes difficult. To make this job easier and quicker, try placing a stake sideways
against the end of the 2×4 and adjust it far enough back so that the thinner
form is supported flush with the 2×4 (FIG. 17-3). The stake could even be
nailed to the 2×4 to ensure accurate placement and support. When thin
form faces are not flush with other forms, lips are created, which are
reflected on hardened concrete. Be sure the tops of both forms are flush,
too.

Driving stakes into the ground next to fence posts encased in concrete might not be possible, and you'll have to come up with a creative
solution. Perhaps a stake can be nailed to a fence post and the form itself
nailed to it (FIG. 17-4). Obviously, metal fence posts cannot be nailed and
you might have to place stakes away from them. In this case, simply use a
2-×-4 block to bridge gaps between forms and stakes. Nail blocks to
forms from their inside faces and place stakes at the butt end of blocks.

Erosion on the sides of slabs and walkways poured along banks and/
or the edge of steep grades is always a concern. Heavy rains and normal
water runoff frequently loosen dirt under slabs and wash it away. A good

17-2 A short stake is used to push out a bow along this walkway form. It will be pulled after concrete has been poured.

17-3 A 1-x-4 form is used to create a wide sweeping curve. To keep its top edge in line with the 2-x-4's top edge, a stake was placed sideways at the end of the 2-x-4 and nailed to it as a means to fully support the 1-x-4 form end in position.

way to prevent this is to dig a 4- to 6-inch-wide trough along affected outer-concrete perimeters once forms have been installed. Trough depth should extend about 2 inches below grade level. Short sections of rebar are driven into the ground along trough paths to serve as anchors for con-

17-4 Concrete poured around the base of fence posts prohibits inserting stakes into the ground next to them. Here, a stake was simply nailed to a fence post to support a form's end.

17-5 This curved walkway section was filled in with dirt to bring it up to a specific grade level. Because of the grade drop off outside of forms, a trough, supported with rebar, was dug out to allow concrete to extend down to actual grade level to add strength to the walkway and prevent erosion.

crete, adding more support and strength to slab edges (FIG. 17-5). Rebar should be driven into the ground at least 6 to 10 inches and rest at least an inch below the slab surface.

Footings generally require rebar installation. Long rebar rods are hung in the center of footings with stakes and tie wire (FIG. 17-6). Lay stakes across footings with one end on top of the form and the other on the ground. Wrap tie wire around rebar and then around a stake once rebar is in position. Tie wire holds rebar suspended in the center of footings. Once footings are filled with concrete, tie wire is cut and stakes are removed; excess wire is simply bent and forced into concrete.

17-6 Wood stakes spanning this footing are used to support rebar rods that run lengthwise inside the footing. The rods are suspended in the middle of the footing by tie wire secured to stakes.

RAIN GUTTER DRAINS

Most houses located in regions that experience seasons of heavy rainfall are equipped with eaves-mounted rain gutters, which channel water to downspouts. Because single downspouts frequently carry water from entire roof sections, a great deal of water can run through them. This water can then, in turn, flood planters or other areas unless a drainage system is in place (FIG. 17-7).

You can direct downspout water runoff anywhere you want by installing sections of 3-inch plastic drainpipe under slabs and walkways (FIG. 17-8). Drainpipe is available at lumberyards and most garden centers. The flexible type is best because it does not require tees or elbows to make connections to downspouts or go around corners (FIG. 17-9). Be sure to dig drainpipe trenches deep enough so that $3^{1}/_{2}$ to 4 inches of concrete can be poured on top of them.

Keep slope in mind while digging trenches. Be certain that pipe runs downhill toward the spot you want water to exit. This could be a curb, central drain, drainage ditch, etc. If necessary, run drainpipe all the way to the front or back of your house to be sure that water does not flood nearby spaces.

17-7 This small planter would have become flooded with every heavy rainfall from the rain gutter downspout had a flexible plastic drainpipe not been provided.

17-8 Flexible plastic drainpipe has been connected to a rain gutter downspout at the corner of this garage.

17-9 Flexible plastic drainpipe is easy to work with.

Drainpipe diameter size is important. Plan to use a 3 inch or bigger pipe to minimize clogging problems. Remember, most of the water runoff will be coming from your roof and a great deal of leaves, twigs, needles, and other debris might be flowing with it. Larger-diameter pipe will allow this type of debris to pass through without clogging. Special adapters are available to connect drainpipes directly to rain gutter downspouts. These work quite well but are not always required. You can simply place pipe under downspout openings for direct drainage.

WATER SUPPLIES

One common concrete oversight is failure to provide a lawn sprinkler water-supply pipe. Getting water from a faucet to a lawn area is quite difficult when a slab or walkway lies between them. Special water jet applicators are available for tunneling under concrete for pipes but why go to the trouble and the expense of renting a tunneling tool when you can easily install a section of sprinkler pipe under a slab before you pour it (FIG. 17-10)? If you do not have any immediate plans for a sprinkler system, you could install a larger-diameter pipe to serve as a conduit sleeve for future access needs (FIG. 17-11).

Sleeves provide access for more than just sprinkler system pipe. They can be used as guides for anything you might want to run under concrete, such as heat pump service lines, drains for outdoor sinks, electrical wires

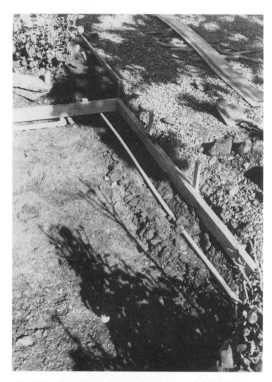

17-10 A lawn sprinkler water supply pipe runs under the end of this formed area and will be covered with concrete.

17-11 A large plastic pipe sleeve runs under the formed concrete pad at the base of this hill. Should electrical wires or other items be needed later they can be easily accessed.

for outdoor service, etc. Sleeves are excellent options and should be considered for almost all outdoor concrete projects.

Placing pipes and other conduit under the edge of concrete as opposed to alongside of it, helps to protect pipes against accidental damage from power lawn edgers or other power tools. In cold regions, pipe can be wrapped in insulation before being covered with a blanket of concrete. Be sure pipe runs downhill toward drain outlets so that water in the pipe can drain out in the wintertime. Otherwise, water inside pipes can freeze and cause pipes to crack.

ELECTRICAL LINES

Have you ever wished that electricity could be provided to a shed or workshop without wires having to be strung from the eaves of your house and over the yard? By laying a section of electrical conduit or plastic pipe under concrete, you can fish wires through at any time. You can even provide electricity to a particular spot in the middle of a slab (FIG. 17-12). You can run pipe, conduit, or wire as needed to the locations desired. Electrical wire should not be connected to any power source until all outer end connections are completely installed and secured.

In preparation for future electrical needs, dead wires can terminate under a small can buried an inch or two under slabs at desired locations. You will have to accurately measure these locations and keep a log in

17-12 Dead electrical wires run through this plastic pipe and into the area under the can. A future fountain adornment will be located above this spot which will require electrical power.

order to locate them months or years later. When you are ready to actually install an electrical device and need to use buried wires, refer to your log and follow the measurements listed. Then, simply break out the inch or two of concrete over the can, remove the can's lid and retrieve wires.

Another important electrical line should be considered close to the edge of hot tubs or spas. This would be an emergency shutoff switch to the hot tub or spa pump. It is dangerous to put any electrical source close to water, so consult a certified electrician or hot tub installer to ensure the electrical switch used and its location are safe. Emergency pump shutoff switches can be lifesavers should anyone get trapped on top of a powerful suction drain. This life-threatening situation has occurred more than once, usually due to the removal of drain covers. Seriously consider this option before pouring concrete decks around hot tubs, spas or whirlpool facilities.

Electrical outlets might be handy at the far end of patio slabs (FIG. 17-13). They can be used for televisions, radios, cooking needs, light fixtures, and other items commonly used during summer barbecues. You might even want to install a natural gas line for your gas-fired barbecue (FIG. 17-14). When installing plastic pipe or conduit under concrete, be sure it is positioned to allow at least $3^{1}/_{2}$ inches of concrete to be poured over it. Cover open ends with heavy-duty duct tape to prevent wet concrete from getting into them during the pour.

If plans call for additional pipe to be attached to a section running under concrete, leave enough pipe sticking out of the slab so that later connections can be easily made. It is better to have too much sticking out than not enough. NOTE: whenever installing electrical or natural gas lines, check with your city or county's building department for guidance.

17-13 A long plastic pipe extension has been taped to conduit below to keep wires out of the way.

17-14 A natural gas line to the left and an electrical supply on the right are taped in preparation of a concrete pour.

Electrical and natural gas connections must be done according to specific building codes in order to be completely safe. Violation of these codes could result in faulty installations which could lead to dangerous fire or electrical hazards. Do yourself and your family a favor by consulting the building department and following their recommendations to ensure safe and long-lasting installations.

FOOTINGS

Unless building codes in your area designate otherwise, footings are generally only required for concrete slabs that will support a structure, such as a room addition, patio cover, shed, etc. Footings are holes or trenches dug in the ground at strategic locations inside formed areas that will be filled with concrete to provide stable support for structures built on top of slabs.

Different geographically located building departments have various footing depth, width, and length specifications for each type of structure. Your local building department can give you these footing dimensions and any other building specification requirements that relate to your job. They are issued at the time you apply for required permits.

Solid footings that run along entire slab perimeters are normally required for room additions, shops, and garages. These footings generally run 12 to 18 inches wide and equally as deep. They are positioned where

17-15 Footings are generally required under exterior walls for room additions and garage structures.

all exterior walls are planned (FIG. 17-15). Rebar rods must also be placed inside footings according to the size and pattern required by building department guidelines. This type of concrete footing provides solid support for exterior walls, and rebar inside it helps to hold all footings together while also adding strength.

Building departments need to inspect footings before concrete is poured to make sure that rebar installation and footing dimensions are correct. Make allowances for this inspection when ordering concrete. Sometimes, building department inspectors are so busy they cannot get out to job sites until a few days after they have been requested.

Patio covers generally require much smaller footings, sometimes referred to as pier footings. These are simply 12-×-12-inch or 18-×-18-inch holes 12 or 18 inches deep located under proposed patio support posts (FIG. 17-16). Pier footing dimensions can vary with each building department so check for exact specifications.

STIRRUPS

Stirrups, also called wet post anchors, are metal brackets that are placed halfway into wet concrete and are used to support patio cover posts. Shaped like the letter *H*, the bottom half is encased in concrete while the top half secures posts. Stirrups are available in sizes to accommodate 4-×-4, 4-×-6, and 6-×-6-inch wood posts (FIG. 17-17). Stirrup placement is critical.

17-16 A 2 × 4 supports this stirrup just for photo purposes. Stirrups are inserted part way in to wet concrete and then used to support patio cover posts. A string is used as a guide so stirrups can be installed straight in line with each other.

17-17 Stirrup footings usually consist of 12- × -12-inch or 18- × -18-inch holes dug 12 to 18 inches deep. The indented legs of this stirrup are designed to go into wet concrete, while the straight sides with pre-drilled holes are made to support wood posts.

If stirrups are not exactly aligned with each other, the patio cover will be compromised. This is because posts work together to support large headers that rest on top of them. Therefore, misaligned stirrups will not allow posts to stand vertically upright and be centered along header lengths (FIG. 17-18).

17-18 This stirrup is securely in position at a corner of a patio slab.

Use nails and string as a guide for stirrup placement. Nails driven into outside form faces designate actual stirrup locations, while a string guide is used to keep all stirrups in line with each other. These guide dimensions must be set up before pours. Nails to support string should be in place so string can be quickly attached and stretched when needed. Place nails on outside form faces to designate stirrup locations. A nail on each side form will hold string stretched tight to determine where stirrup edges should be so they are in line.

For the most part, building codes usually allow 10-foot spacing between stirrups with extensions of no more than 3 feet overhang off the ends. If you were going to erect a patio cover measuring 36 feet across the back, for example, you would need four stirrups. From a side form, one would be at 3 feet, the next at 13 feet, then at 23 feet, and finally at 33 feet. The 3-foot overhang at the opposite end would bring the total to 36 feet. This is only a general stirrup placement rule of thumb, with precise spacing designated by overall patio cover dimensions and the type and size lumber used in its construction (FIG. 17-19).

Although stirrup placement is officially guided by building department regulations, you should keep in mind that an edger tool will have to be used along end form edges. Therefore, try to install stirrups 6 to 12 inches in from end form edges to allow plenty of working room for the edger tool.

17-19 Black plastic flower pots have been placed upside down over installed stirrups to protect anybody that might trip on them.

Stirrups are only used to stabilize fence posts, patio support posts, and open carport posts. If your concrete slab is planned as a floor for a room addition or garage, footings and anchors will need to support entire walls. For these structures, you will need to use J-bolt anchors (FIG. 17-20).

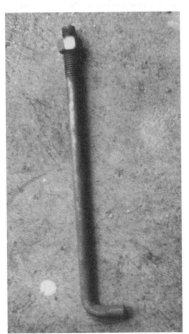

17-20 A typical J-bolt used to secure wall structures to concrete slabs.

J-bolts are placed in wet concrete after the first bull float, as determined by designated nails partially driven into outer form faces. When concrete has cured, a 2-×-4 or 2-×-6 mud sill is secured to J-bolts. Walls are then erected on top of the 2-×-4 or 2-×-6 mud sill. J-bolts are generally placed about 1 foot in from each corner and then about 4 feet apart. Follow your building department's guidelines for exact spacing and placement. Do not put J-bolts in spaces that are planned as doorways.

When inserting J-bolts, use a small, 2-×-4 board as a depth guide. Push J-bolts into concrete in line with nail locators and string guides, making sure enough of the bolt sticks up to secure a 2×4. You can either allow J-bolts to stick up 1/2 to 1 inch above the 2-×-4 guide or make the bolt top flush with the 2-×-4 top. If flush, you will have to drill countersink holes in mud sills to allow room for washers and nuts. Refer to your building department's guidelines for precise instructions.

REINFORCING WIRE

Reinforcing wire, also called hog wire or 6-×-6-×-10 wire, is 10-gauge steel wire made into 6-×-6-inch squares (FIG. 17-21). This wire comes in rolls 7 feet wide and 50 to 100 feet long. You can buy as many linear feet as needed, all of which will be 7 feet wide. Hog wire is available through lumberyards and most concrete batch plants.

Hog wire adds some strength to concrete but is primarily designed to keep concrete cracks from separating. This is especially important to keep

17-21 Hog wire, or wire mesh, helps prevent cracks.

cracks flush, as one side moving upward will definitely become a major tripping hazard (FIG. 17-22). More of a luxury than a real necessity for patio slabs and walkways, hog wire is a real plus for driveways and other slabs that will have to support a lot of weight.

17-22 Concrete pourers must periodically reach down during pouring activities and pull wire up and into the concrete.

Placing wire will require bolt cutters and a helper. Bolt cutters are needed to cut wire to length and a helper makes moving large sections of wire a lot easier than doing it alone. Keep the 7-foot width in mind while cutting sections to fit inside your slab to keep overlapping at a minimum. If your driveway measures 12 feet by 28 feet, for example, four 12-foot sections (48 linear feet total) could be placed side by side for complete coverage as opposed to two 28-foot sections (56 linear feet total) placed lengthwise with a 2-foot overlap.

Lay wire down so that its natural curl faces toward the ground. Hog wire tends to roll back on itself if laid down with the curl up. Simply unroll the prescribed number of feet needed, cut, and then flip it over for placement (FIG. 17-23). The ends will dig into the ground and you will have to force the middle to lay flat. Smack sections with a shovel to help force them down. Wire needs to lay flat below concrete slab surfaces. If pieces are noticed rising through the surface during finishing endeavors, you will have to cut them out.

Although hog wire must lay flat under concrete, it should not be flat against the ground (FIG. 17-24). You want it somewhere in the middle of concrete so that all sections are completely encased. To ensure that, reach down periodically and pull wire off the ground while pouring concrete (FIG. 17-25).

17-23 Hog wire tends to roll up on itself unless placed on the ground with its natural curl facing down.

17-24 Sections of hog wire can be cut out to accommodate other items, such as a rain gutter downspout drainpipe.

Some concrete finishers like to place rocks under hog wire at different points to make sure concrete will completely encase it. On large pours, though, this practice has been found to be a nuisance, as wire sticking up an inch or two off the ground presents itself as a continual tripping hazard. When using hog wire, just plan to reach down every few feet and pull it up into concrete.

17-25 The man controlling a concrete truck's unloading chute is bending over to pull up a section of hog wire. Some concrete finishers like to put rocks or bricks under wire to keep it suspended.

CONCRETE PUMPING PREPARATIONS

Concrete pump operators can be found in the yellow pages or through a concrete company. Concrete pump hose generally comes in 20- to 25-foot sections. The first section is placed at the farthest end of the job site and subsequent sections laid down until they finally reach the pump (FIG. 17-26).

All concrete pumping jobs should have one helper designated strictly to pull and maneuver the hose out of the way once pumping begins. Kinks must be avoided at all costs. A kinked hose on a positive displacement pump causes a relief valve to operate, which takes around 20 minutes and a lot of work to reset. If stringers are installed in forming designs, make sure that they are securely supported and braced with extra stakes if a hose must be laid over them (FIG. 17-27).

Concrete pumps are primed with their hoppers full of water and cement powder and maybe a friction-reducing soap agent to make a slurry. This is the first thing to exit through the end of the hose, and you must have a place for it to go outside of forms—it should not be dumped inside formed areas. Have the pump hose stretched out far enough to reach the furthest slab ends and then direct it to where you'll want slurry and water solutions to drain (FIG. 17-28).

Some concrete pump operators will man the end of the hose for you and rely on a remote switch to turn the pump on and off. Other operators stay with the pump at all times, and you must supply someone to man the end of the hose. This can be discussed when you talk to a pump operator.

17-26 A connection is being made between two lengths of small-diameter, concrete line-pump hose. Both ends of the hose are alike. The one on the right has had a rubber gasket placed over it. Once they are butted together, a clamp coupling holds them together.

17-27 Forms and stringers must be braced and supported with extra stakes when a pumping hose is laid over them.

17-28 This concrete hose has been stretched so that it will reach the furthest section of the job but has been doubled back to an area where priming water and cement slurry can be flushed without causing any problems.

Once the slurry and water solution has drained out of the hose and concrete starts to flow, move the hose end to where you want to start your pour; generally the area furthest away from the pump. Hold the hose end close to the ground, about 3 to 6 inches above grade. This greatly reduces concrete splatter as mud hits the ground and also helps you to correctly gauge concrete depth with form tops. You'll know when concrete is about to flow out of the hose when you hear a distinctive "swishing" sound of moving pea gravel and sand and the hose starts to get quite heavy.

If a pump operator volunteers to man the hose, let him! His experience and expertise should help your job go smoothly. If not, pay close attention to how fast concrete exits hose and keep the end moving to prevent unnecessary concrete build-up. If pouring gets ahead of screeders, signal the pump operator to shut down (FIG. 17-29). Give screeders a hand, if necessary, to help them catch up. When ready, signal the pump operator to start up again.

When job sites are located in backyards and pumps remain out front, you might have to have a runner available to relay messages back and forth from the job to the pump. If a remote switch is available to turn the pump on and off, use it. By all means, clearly establish and use a means of open communication between you and the pump operator.

17-29 At least one person must be put in charge of nothing except pulling and maneuvering concrete pump hose.

PREPARING FOR THE SECOND LOAD

Novice concrete finishers without any experienced help on site should keep their first concrete jobs under 4 yards maximum. Should jobs become sufficiently big enough to require more than one truck full of concrete, several things must be taken care of in preparation for subsequent deliveries.

Enough helpers must be on hand so that at least two of them can begin tamping and bull floating while the second load is being poured. By the time this second load is on the ground, the tamper should be finished and ready to work on fresh concrete with the bull float operator close behind. If not, a first load could flash by the time a second load is poured, and you will have a tough time successfully tamping and bull floating. This will essentially put you behind on your job and you'll have to work extra hard and fast to catch up.

If the second truck is ready to go as soon as the first one empties, rake the last concrete edge to a gentle slope. Pour new concrete from the second truck on top of this slope and screed normally. If you have to wait a while for a second truck, use a rake to make a 4 to 6-inch-wide step at the end of the concrete on the ground about 1 inch deep. This way, a fresh load of concrete poured 4 inches deep will not butt up to a previously poured 4-inch-thick slab. In that case, you'll have formed a cold joint; a situation guaranteed to eventually present cracks. Rather, new concrete will have a wet and even concrete base of about 3 inches to rest on and the 1-inch-thick concrete on top will mix well with the other surface after tamping (FIG. 17-30).

As important as it is to have extra helpers available to work on the first pour while unloading the second truck, you must be sure the base is sufficiently wet down before the second truck starts dumping concrete (FIG. 17-31). Your second pour will take place a little later than the first and chances are some moisture will have evaporated. Wet bases with a garden hose.

CHEMICAL ADDITIVES

Many professional concrete finishers have successfully used chemical additives to help concrete set up faster during cold, wet weather or to help concrete remain wetter during hot dry weather. Chemicals used to speed up or slow down flash times are put in at concrete batch plants.

Calcium Chloride is one product that is supposed to help concrete dry out and flash faster. Especially during winter months when days are short and the weather cold, wet, and miserable, finishers have dispatchers add specific amounts of "C-C" to help moisture evaporate quicker so that finishers can hand trowel surfaces sooner and complete jobs quicker. Your need to use such a chemical product is debatable. Rely on advice from the concrete dispatcher for your regional area.

Many concrete finishers have said they poured in cups of sugar to loads of concrete to prevent mixes from setting up too fast in hot weather.

17-30 This fresh concrete edge has been properly raked and stepped in preparation for a second concrete load delivery.

17-31 Base material, or dirt, must be thoroughly wet down before any concrete pour.

Others profess to have used molasses for the same reason. Some say they like to use a retarder to keep surface cream softer for longer periods to assist in exposing aggregate slabs.

Whatever the reason, discuss the use of retarders with your concrete dispatcher before attempting to add anything like sugar or molasses to any concrete load. If the scorching weather in your area is sure to flash a slab in no time, rely on a professional retarder product. Have the batch plant operator add this chemical to your concrete load as prescribed by the product manufacturer and in line with his or her experience mixing concrete in your area and its prevalent weather conditions.

Chapter **18**

Special pouring techniques and custom finishes

*P*ouring and finishing flat work concrete is a straightforward process. Areas have to be formed to hold concrete in particular shapes, concrete has to be delivered from a truck to those forms, and then it must be flattened and smoothed. This basic process becomes a bit more complex, however, when special forming designs, limited concrete delivery access, and custom finishes are introduced.

As you become more proficient in concrete flat work, you will most likely prefer to pour custom work in lieu of standard broom-finished slabs and walkways. From the initial planning stage through all phases of your job, however, you must keep the basics in mind so that each task is completed proficiently and in a timely manner. Be as inventive as necessary, like removing fence sections to allow controlled bull float and fresno work, hiring a concrete pump company for limited access jobs, and using large sheets of plastic to build tents over jobs to protect them against inclement weather while you complete finishing maneuvers.

CONCRETE PUMPS

Concrete line pumps are great machines (FIG. 18-1). They do a splendid job of delivering six-sack pea gravel mixes to formed areas, even though pulling a hose can be a cumbersome chore. If your custom job features a lot of concrete spread over a wide area, like extensive walkways and patio slabs, or the amount of maneuvering room for a concrete hose is limited, you might consider using a "boom truck" concrete pump.

Boom trucks are equipped with a large concrete pump and an articulating boom. The larger pump is capable of pumping concrete with up to 1-inch diameter aggregate and can easily go over most houses to reach

263

18-1 Concrete truck drivers control the flow of concrete into pump hoppers and remove large rocks that inadvertently get caught on hopper screens.

backyards. In addition, the hose does not have to be dragged from place to place. A section of hose dangles from the end of booms to make actual concrete placement more manageable.

Whenever pulling or maneuvering concrete pump hoses, especially the smaller-diameter line-pump types, be extra alert to avoid kinks of any kind. This is a prevalent concern on custom jobs with curved forms or lots of stringers. Helpers should also be alert to avoid form damage while dragging a hose. Again, if your custom concrete flat work job employs a lot of unique curves, bends, stringers, or a widespread design covering a lot of surface area, you might be best off with the services of a boom truck to alleviate the need for extensive hose dragging and maneuvering.

SPECIAL CONCRETE POURS

Pouring concrete against benderboard or forms taller than 4 inches requires a slow buildup rather than an immediate full pour. Start out by

18-2 Gently pour concrete against benderboard forms to reduce the risk of pushing thin forms out of position.

pouring an inch or so of concrete against these special forms (FIG. 18-2). This lessens the possibility of thin benderboard forms bowing under the sudden onset of a heavy concrete load. Same for tall forms. A rapid concrete pour of 6 inches or more against these forms could cause them to collapse or push them out of plumb (FIG. 18-3).

For these types of jobs, pour an inch-high load of concrete for about 6 to 10 feet. Then go back and fill in another inch-high load. Once that level matches the surrounding grade, pour normally.

Steps present the same concern. Too much concrete poured into forms too fast can cause step-face forms to bow or break loose (FIG. 18-4). This is especially true for custom curved steps formed with benderboard. In this case, consider filling formed spaces with a shovel. Not only will it minimize the chance of form movement or a blow out, it also accurately places concrete correctly with each shovel full.

Pool decks have unique concrete pouring concerns. Of utmost importance, of course, is not falling into the pool, which has happened more than once. Pool decks must be formed with a definite slope away from water to prevent dirty water or rain runoff from entering the pool. Screeding must be controlled to prevent excess concrete from falling into

18-3 A sudden, full-force concrete pour against these benderboard forms would surely knock them out of position.

18-4 A shovel is used to carefully place concrete in this benderboard-formed area.

18-5 It is best to hand float concrete around inner pool deck areas before screeding to minimize the amount of concrete that could drop into the pool.

the pool area (FIG. 18-5). You might find that hand floating concrete up to pool edges before screeding works best.

Whenever wheelbarrowing concrete into pool deck areas, empty them in such a way that concrete is not directed toward the water. Consider having a helper hold a piece of plywood in front of wheelbarrows as they are emptied to hold back mud and prevent it from dumping into the pool.

Pool decks must be equipped with expansion joints just like any other slab (FIG. 18-6). In lieu of stringers, felt, or benderboard, however, plan on using materials designed especially for pool decking. Referred to by many as *pool cap*, these plastic strips incorporate a series of underlying channels that help to funnel water away from pool areas toward outer deck perimeters. They are installed perpendicular to pools and feature an adhesive protective strip along top faces that is removed after concrete has cured. Pool cap is available through concrete equipment supply firms, pool supply stores, some ready-mix concrete companies, and lumberyards.

Blocks used to help align forms during installation can be left inside formed areas until concrete replaces them (FIG. 18-7), such as when benderboard forms are used to outline certain jobs. Blocks support forms while concrete is poured against one side. Then, as concrete is poured against the other side of the form, blocks are gently removed while concrete is pushed into the open areas vacated by blocks (FIG. 18-8). In some cases, it might be best to place concrete with a shovel to guarantee forms are not moved about by the force of heavy pouring from a truck chute or wheelbarrow (FIG. 18-9). When the space is completely filled with concrete, jab a shovel into the mix a number of times to help blend all material together to prevent cold joints from forming.

18-6 Expansion joints must be provided around pool decking. Here, 2-x-4 stringers are used. The top of this stringer is being cleaned off so that a clear stringer edge is visible for future edging.

18-7 Blocks are kept in place to support a benderboard outline form during a pour and removed when entire areas are filled with concrete.

18-8 Blocks are carefully raised as concrete is forced under them to fill the void. This ensures benderboard is constantly supported; either by a block or by concrete.

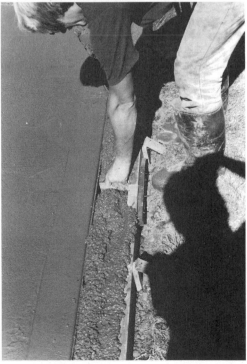

18-9 Concrete has completely filled this section before the block is removed. Notice how straight the benderboard outline form is.

18-10 For intricate concrete placement, consider pouring, tamping, and floating a slab's main body before spending a lot of time with the special work.

Custom concrete jobs with a mix of special facets like steps, color, outlining, and curves require extra time and attention for detailed concrete placement. It might be best to pour, screed, tamp, and bull float the slab's main body first so sufficient time can be afforded for intricate items without causing unnecessary delays in the overall job (FIG. 18-10).

For large pours requiring two or more trucks of concrete delivery, you will have to properly prepare the first load's last edge in preparation for the second delivery. You can rake off the concrete's edge to a slope or prepare it into a step arrangement on top of which the second load will be poured—this is done to prevent cold joints.

Another way to prevent cold joints is to calculate just how far the first load of concrete will stretch and then plan an expansion joint at that point. For example, let's say you are going to pour a 12-yard driveway. Because the largest concrete truck can only carry 10 yards at one time, the job will require two loads. Why not request two trucks with 6 yards each?

You can place a temporary form at the point where the first 6 yards will end. A section of expansion joint felt should be placed along the inside face of that form so the first load of concrete is poured against it. Pour to that form, screed, tamp, and bull float concrete before the second truck arrives and then slowly remove the form as concrete is poured against it from the second truck. Not only will the possibility of a cold joint be avoided, the insertion of an expansion joint in the middle of your slab will have been easily accomplished.

CUSTOM FINISHES

Outdoor concrete flat work should never be left perfectly smooth. Hand trowel or power trowel finishes alone leave concrete surfaces blemish-free but also allow them to be extremely slick and slippery when wet. To provide traction for wet flat work surfaces, concrete finishers apply broom finishes, custom finishes, or a combination of broom and custom applications (FIG. 18-11).

18-11 A combination of stringers and open planters help to make this rock salt finish look great.

You will pour, screed, tamp, bull float, fresno, edge, seam, and hand trowel concrete flat work the same for all custom finishes except exposed aggregate applications. Exposed aggregate finishes are not tamped. This finish requires aggregate to remain close to a slab's top surface so excessive amounts of cream do not have to be washed away to expose aggregate (FIG. 18-12). Other than that, you will finish jobs regularly until it is time to effect custom finishes.

Custom concrete finishes are not limited to simply etching, rock salt, or exposed aggregate designs. Many innovative finishers have developed unique ways of customizing slabs to fit within certain landscape and architectural schemes. Some concrete equipment companies sell certain "stamps," which are used to make slabs appear as if they were made of brick.

Stamps are tools that look very much like tamps except for their bottom bases. Instead of a screen mesh for pushing aggregate down in wet concrete, stamps feature metal bases shaped into specific brick patterns. After concrete is finished, stamps are placed on surfaces and pounded down with a hammer to make brick pattern impressions. This is a labor-intensive operation, as stamps only cover a few feet at a time. You can locate stamps through concrete equipment supply outlets or concrete company dispatchers.

Etching finishes

To apply an etching design finish, you'll need a 1-inch-wide wire brush that is in good condition with long stout bristles (FIG. 18-13). Before etch-

18-12 Exposed aggregate finishes require that a top layer of cream be removed to expose the aggregate underneath. Here, aggregate is starting to be exposed through water spray and broom activity. Sand, silt, and concrete residue accumulate on the right side of the picture.

18-13 A 1-inch-wide wire brush was used to create this etching finish.

ing commences, slabs must be hand trowel finished, broomed, and set up enough so knee boards will not leave any marks or impressions. Slabs should be poured early in the morning so that there is plenty of sunlight left after concrete has hardened to illuminate the work area for etching applications.

Effecting crisp etching lines the day after a pour is very difficult, as slabs have generally hardened too much and brushes cannot sufficiently etch surfaces. Because etching must be done on the day of the pour, finishers prefer to etch along forms first to let main slab bodies have more time to set up.

Etching marks are made by drawing a 1-inch-wide wire brush across concrete surfaces in a series of freehand fashions. Although there is no set pattern to this type of finish, most professionals prefer to round off all formed corners and points where etching lines intersect.

Two wire brush passes is often enough to effect crisp etching lines. The brush is always pulled toward the operator, never pushed out in a forward motion. You will want etching lines to be uniform insofar as they roughen the concrete surface to make it look like mortar between bricks. Etching lines need only be deep enough to effect a rough mortarlike texture. When one pass fails to produce satisfactory results, duplicate with a second pass (FIG. 18-14).

After you have outlined a slab by etching along forms, get your knee boards and go out to where concrete was first poured. This is the place where the slab surface should be hardest and most set up because it has been on the ground longest. If a part of the slab has been shaded the

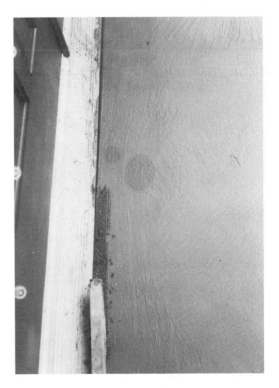

18-14 A 1-inch wire brush starts to make an initial etching line. Two passes are usually needed to etch concrete completely and present crisp lines.

entire day and is still wet, save it for last so that it can set up enough to resist any knee board impressions.

Your freehand design starts about 1 to 2 feet out from a corner with a wavy arc going from one form to about 1 to 2 feet out on the other form (FIG. 18-15). Round off intersections where etching lines meet forms. From this point, you will continue to make wavy arc lines that intersect with one another to form "stepping stones," all about the same size. As you progress, you'll get a feel for the design and probably begin to notice somewhat of a pattern developing; like, circles, kidney shapes, or what have you. In this case, try to continue along the same so-called pattern to maintain a degree of uniformity.

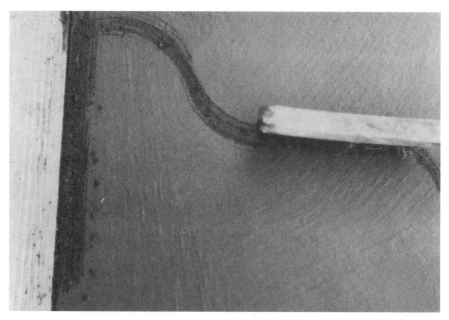

18-15 This is an example of how a freehand etching line might be started. All you have to do is draw your brush back in any direction desired. Two passes will be needed for this line and the intersection of both lines must be rounded off.

Whether or not any sort of a freehand pattern takes shape, try to at least keep etching lines separated enough to allow "stepping stones" to remain about the same size. Overall, though, a freehand design's only limit is your imagination.

Note: small crumbs of concrete will accumulate on slabs that are etched. Most of this residue is swept off cured slabs with a soft push broom when etching is completed. Other than that, simply wash crumbs off with water a few days later when concrete is hard enough to allow you to walk on it without leaving any marks or impressions.

Rock salt finish

A rock salt finish incorporates hundreds, if not thousands, of small holes in concrete slab surfaces. Although eye-appealing, these holes tend to trap dirt and debris and must be washed off with water to get clean (FIG. 18-16).

18-16 Rock salt holes appear to be more concentrated in some areas than others. Make sure granules are spread evenly throughout the slab surface.

Morton Brand "Coarse" rock salt is a popular size for a lot of concrete finishers; granules are about the size of a garden pea. Rock salt comes in different sizes from large rocks (which make large holes) to very small granules (tiny holes) and you have to decide which size would look best on your slab. One large bag, about 20 pounds, will generally cover around 300 square feet, approximately 4 yards of concrete flat work.

This finish requires that concrete be hand trowelled and broomed as soon as possible. You will need a light touch to finish off slab surfaces that are still a little on the wet side. Concrete must be slightly wet in order to allow rock salt granules to penetrate surfaces intact. Gauge a slab's hardness by pressing your finger against it. Ideally, this will leave an impression slightly heavier than a simple fingerprint.

Broom lines should be very easy to put on slightly wet concrete slab surfaces. If need be, apply upward pressure to extension poles so broom bristles do not dig into concrete too heavy. After a slab is broomed, toss on handfuls of rock salt as uniformly as possible. Optimum coverage would be an evenly dispersed coat of rock salt spaced no more than an inch apart (FIG. 18-17). Uniform rock salt granule coverage is a key factor in producing an attractive rock salt custom finish. Areas with too much salt will look awkward in comparison to those with extra light coverage (FIG. 18-18).

Rock salt will not make impressions in concrete alone, you have to pound granules in with a metal trowel or hand float. On narrow walkways and around slab perimeters, reach into slab areas while kneeling outside of forms to pound salt down into the concrete surface (FIG. 18-19). Keep

18-17 Forms are not pulled until all rock salt has been pounded in; evidenced here with granules smashed all around the slab.

18-18 Large-sized rock salt granules were used on this slab, resulting in rather large holes.

18-19 The tool being supported here is a homemade rock salt roller. It consists of a large plastic pipe filled with concrete and equipped with an adapter and pole. Its weight works great for forcing down rock salt on soft slab surfaces.

your metal hand trowel or float flat to maximize its pounding effectiveness and avoid scarring concrete with an edge or corner of the tool. Repeat efforts until salt has made a definite impression on surfaces.

Use knee boards to leap frog around slabs as you did while trowel finishing. Reach as far out as possible to pound salt and minimize knee board movement (FIG. 18-20). Once all salt granules have been pounded down, the job is done. Should you notice an area salted too lightly, toss on a handful or two of granules and then pound them in (FIG. 18-21).

Rock salt is washed off or dissolved from slabs as you hose down concrete to maintain moisture. After 3 or 4 days, you should be able to safely walk out on slabs and completely wash off all salt residue.

Exposed aggregate

Exposed aggregate finishes require that layers of cream be washed off surfaces to allow aggregate to be clearly visible. Slabs and walkways are finished exactly the same as with any other except that they should not be tamped. This keeps aggregate close to the surface and minimizes the

18-20 A knee board and a couple of 2-×-6 boards are used to access the slab and pound salt into surfaces. They are supported by stringers to minimize the amount of pressure on the actual slab.

18-21 Rock salt coverage is even and all of the finishing work was completed before forms were pulled.

amount of cream that has to be washed away. Be sure to accomplish good screed work, as well as bull float, fresno, edge, and seam chores. Hand trowel work is not critical but should be instituted to maintain smooth and even slab surfaces free of significant fresno or trowel lines.

For all intents and purposes, exposed aggregate slabs should look like any other slab right up until the point you decide to start washing it off. The reason for this is to maintain a uniform aggregate and cream height throughout all slab areas so that when washing begins all surface areas will feature a uniform texture and appearance.

Concrete must be hard enough to walk on without leaving impressions or dents before washing begins. Although you'll want to rinse off surface cream, best results are afforded when underlying concrete is not disturbed or weakened. Test your slab's hardness by simply stepping on a part of it. If an impression or dent ensues, it's too early. Walk on fresh slabs with a light touch and do not twist or grind your feet against the surface. Remember, concrete is still vulnerable even though it will support your weight when you walk flat and evenly.

Start the process by sprinkling water on the slab. Use just a spray of water and never a straight or direct stream. Straight streams of water dig into concrete surfaces and gouge out small chunks or create little troughs. Employ a soft bristled push broom to loosen up the top layer of cream after it has been wet down with a water spray (FIG. 18-22).

Go slow and gently. This is not a quick, one-shot process. Most slabs have to be washed off two to three times and none are ever satisfactorily completed with just one water and broom application. Don't try to remove a lot of cream all at once. Vigorous water and broom applications tend to dig too deep into surfaces and will eventually result in an uneven exposed aggregate surface. With the first washing, you should be satisfied

18-22 Two finishers work together with water and a soft-bristled push broom to remove a top layer of cream while effecting an exposed aggregate finish.

to break loose the top cream layer and bring up just shallow accumulations of sand and cream.

Push and pull the broom across small areas at a time until you can barely recognize aggregate, then move on to another section. Large accumulations of loosened material are swept off the side of slabs with a broom (FIG. 18-23). It may take some time to effectively maneuver the hose and broom across your slab. Again, don't get in a hurry and do not expect this very first washing to produce final results. When the first washing is complete, carefully inspect concrete to be sure the surface has not softened up. Don't worry about cream, be concerned about loosening aggregate underneath instead. Optimally, very little aggregate should be washed away. The only residue you really want to get rid of is cream and sand.

The second washing will produce much more visually pleasing results. With the top layer of cream loosened up, a second water and broom application should clearly expose aggregate (FIG. 18-24). If concrete is stable and hard, you can change the water spray pattern to a closer stream for more effective rinsing results. Keep in mind that you will continue to wash the slab until the only visible residue is clear water. So don't be overly aggressive. Try to maintain an even and uniform spray and broom application to develop a preferred depth where the top portion of aggregate is clearly exposed yet not so deep that individual rocks are sticking up with little material surrounding them to keep them in place (FIG. 18-25).

As accumulations of cream and sand amass, use a broom to sweep them off the side. Refrain from using excessive water spray to continually move accumulations of concrete residue ahead and out of the way. This

18-23 As sand and concrete residue build up on top of slabs, use a broom to sweep them off the side. If water alone is used, the slab surface could be gouged.

18-24   Once an initial water spray and broom effort have loosened top-cream layers, a second rinsing will start to reveal a rich-looking exposed finish.

18-25   This slab was wet down with a garden hose before this finisher started using a soft-bristled broom to loosen the top layer of cream for an aggregate finish.

will only cause more water pressure to be applied to the surface, running the risks of gouging out small sections of concrete in the process (FIG. 18-26).

If your slab has set up so hard that it resists a soft-bristled broom, try using a stiff-bristled one. Work gently at first to see how effective the stiffer bristles are. If they seem ineffective, have a helper assist you in applying water and broom strokes at the same time on the same spots (FIG. 18-27). Too much broom penetration on softer slabs, however, causes aggregate to loosen, which could lead to small holes or pockets in the surface (FIG. 18-28).

18-26 If water alone does not seem to be cleaning slabs off sufficiently to produce an exposed finish, have a helper work with you and use a broom for extra cream removal.

18-27 As this slab became harder, a stiff-bristled broom was needed to break loose and remove cream.

18-28 Exposing this driveway slab is time-consuming. Already, it has been rinsed two times. The section in the shade was too wet earlier and is now being exposed with light water and a soft broom.

Use short, even broom strokes to simply move a little bit of cream each time. It is much better to make two or three controlled light sweeps than one extra-strong scrape (FIG. 18-29). Apply water streams and broom strokes as uniformly as possible so that an even amount of cream is removed with each pass. This will help immensely to maintain an evenly textured exposed aggregate finish.

Constantly keep your water nozzle moving. Concentrating the spray on one spot at a time will remove too much material and either gouge out a trough or cause material in that spot to loosen more than necessary, becoming susceptible to holes or pockets. As with the broom, two or three light water applications are much safer and more effective than a single heavy dose (FIG. 18-30).

The second washing will produce less concrete residue than the first. In fact, water spray will be much more effective the second time around (FIG. 18-31). Aggregate will become more pronounced and your slab will start to look pretty good. Continue to wash in the same manner, although you will have to rinse the slab off at least one more time (FIG. 18-32).

On a sunny day with the temperature rising to about 75 degrees, you can expect to apply a third washing about nine hours after concrete has been poured. For a slab that was poured at 6:30 AM, this would be around 3:30 PM. This is an important consideration. Should your pour take place in winter, you might not have these warm temperatures to help concrete

18-29 This broom is used in all directions to break loose a cream layer.

18-30 Water will wash off concrete cream from fresh slab surfaces. Insist on applying two to three light water spray applications as opposed to a single heavy one.

18-31 A section of concrete that had been shaded all day still appears somewhat wet.

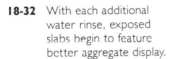

18-32 With each additional water rinse, exposed slabs begin to feature better aggregate display.

set up. In this case, seriously consider the addition of an accelerator and a mix with hot water to help speed the concrete curing process. Check with your concrete dispatcher.

After the second wash, you might be inclined to call the job "good" and stop there. Aggregate should be clearly exposed and because it is wet from water spray, it will look dark and rich. However, there can sometimes be a very fine film of concrete residue covering the slab. Have you ever noticed how some exposed aggregate finishes look great and have a dark rich color while others have sort of a white film over them and look kind of a uniformed light gray color? The difference is the number of washings each received.

The light colored, and generally dull looking, slab was probably only washed off one time. As a result, a thin layer of cement and cement by-products has formed over the surface. The dark and rich-looking exposed aggregate slab was washed off as many times as it took to remove all traces of any surface film to clearly expose aggregate underneath (FIG. 18-33).

18-33   The third water rinse for this exposed aggregate finish is proving fruitful. Notice that water runoff is clear, an indication that surface cream has been removed.

The general rule of thumb for exposed aggregate finishes is: continue to gently wash off slabs until water is clear and is no longer removing any concrete residue (FIG. 18-34).

Very small patches of concrete tucked away in remote corners and covered by shade all day might not stand up to a lot of water and broom activity. For those, use a trickle of water and very light broom applications (FIG. 18-35). You might have to wait up to a half hour between washings to be sure that underlying concrete is not unduly weakened by premature washing activity.

To ensure a long-lasting, deep rich color to exposed aggregate finishes, consider spraying concrete sealer on top of completed slabs. Sealers do an excellent job of holding color in and even make some aggregate finishes look better. Sealers are available at concrete batch plants and some lumberyards.

18-34 This concrete driveway has been successfully exposed. The sidewalk area closest to the curb was not exposed. To keep the majority of water off of it, a hose was laid across the slab just in front of the sidewalk to divert it.

18-35 This custom slab features a colored concrete border around a center section that is being exposed.

APPLYING COLOR

You can order concrete in an assortment of colors direct from a ready-mix plant, which will also supply a color chart. As concrete is mixed and fed into a concrete truck, color dye is added in prescribed weight dosages to arrive at specific tints. This results in your entire concrete load being dyed the chosen color.

You can also apply color dye to slabs yourself. Color dust, or dye, is sold in one-pound bags with one pound able to tint about 20 square feet of concrete. Prices vary between concrete companies and certain colors are much more expensive than others.

Color dust is sprinkled on slabs after tamping and just before using a bull float (FIG. 18-36). It is actually the bull float that helps to spread dust around to a uniform color, provided dust application was somewhat uniform (FIG. 18-37). On small jobs with easy access to slab areas, this is no problem (FIG. 18-38). Larger jobs may be more labor-intensive.

When you pour large jobs, plan to work in about 4-feet increments. Tamp concrete on the ground and then sprinkle color dust on it. Use a bull float to smooth the surface and work in color. When that section is complete, start pouring again until 4 feet is on the ground, then repeat the process. This is a time-consuming chore and best accomplished with plenty of helpers. You should tell the concrete dispatcher about your plans to color dust slabs because there might be some standby time accumulation. This courtesy will help dispatchers to schedule other deliveries around the extra time you will have his delivery truck tied up.

Concrete color dye, or dust, is very potent. It will stain everything it comes in contact with, including tools. If your job is a combination of color-dusted concrete and plain concrete or more than just one color, be sure to have enough tools on hand so that a separate set is designated for

18-36 Concrete color dust is spread out on top of a custom border design.

18-37 The top step border has been color dusted, hand floated and edged. Here, the lower border has already been dusted and the finisher is using a hand float to work in, and blend, color over the entire area.

18-38 A hand float is used to gently work color into concrete.

each color. Using tools stained with red color dust might not work too well on concrete dusted with light brown or vice versa.

Applying color dust to step faces is tricky (FIG. 18-39). Unless yours is just a single small step, you will be much better off having an entire load of concrete delivered in the color desired. To do a good job of color dusting step faces, you will have to have some extra concrete cream on hand to mix with dust and then apply to step faces after forms are pulled (FIG.

18-39 Color dust is carefully sprinkled down a step-face form to tint concrete on the step face.

18-40). Use a metal hand trowel to work up some moisture on the step face and then trowel on a layer of colored cream. Once the correct and uniform color application has been completed, finish the step face normally.

Some early hand trowel work might be required to help blend dusted areas in with each other, but generally, after color dust has been applied and floated in, dusted slabs are treated exactly the same as any other. Subsequent bull float, fresno, edging, and other work is unchanged (FIG. 18-41).

To maintain rich, dark and glossy colored concrete, you have to apply a sealer to the surface after it has cured. Sometimes referred to as wax, sealer products do an excellent job of protecting colored concrete against fading and becoming dull as a result of harsh weather and constant exposure to ultraviolet sun rays. Check with your concrete dispatcher to determine which sealer is most appropriate for the color you have chosen.

BRICK INSERTS

As long as space has been provided in formed areas, bricks of just about any style with just about any type of finish can be inserted into slabs (FIG. 18-42). Areas left open for bricks must be wide enough to accommodate the mortar required to keep bricks in place (FIG. 18-43). Don't forget that 2-×-4 forms are 1¹/₂ inch wide and the space they occupy during a pour will become empty once they are removed. Neglecting to account for

18-40 After color dust is sprinkled between the step-face form and concrete, the form is gently tapped to vibrate concrete cream against the form, mixing color as much as possible.

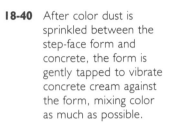

18-41 Once color dust is applied and blended in, normal concrete finishing activities continue. Here, a finishing trowel is used to smooth the surface.

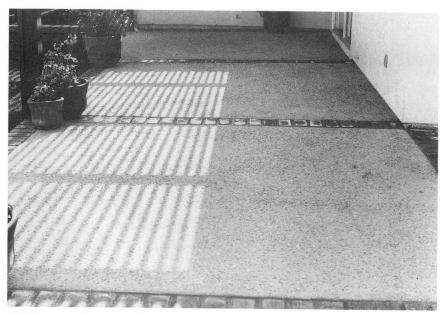

18-42 Bricks are inserted into slab areas after concrete has been poured, finished, and cured.

18-43 Bricks can be installed in any pattern desired as long as gaps are wide enough for their installation.

two, 2-×-4 form widths for a brick insert will result in the gap being an extra 3 inches wide when forms are removed.

You can design any type of brick insert shape desired (FIG. 18-44). Just be sure you have given plenty of forethought into the forming process and measurements are accurate. You must also ensure enough forms and stakes are on hand to support additional forming needs.

Bricks are inserted after concrete has cured. A sand base is first laid down along gaps. Use as much sand as necessary to enable brick faces to sit flush with a slab surface. Use a small 2-×-4 board to span the gap, like a screed board, and adjust bricks to its bottom edge. This is a handy brick height gauge. After bricks have been properly situated and adjusted in place, sprinkle about a half inch of sand around them to help maintain their position and keep them from moving during subsequent operations.

Dry brick mortar is poured around bricks. After all of the slab or walkway's brick inserts have been properly positioned and secured with a half inch of sand around them, you are ready to apply mortar. Tear a small hole in a corner of the mortar bag and pour the fine-powdered material into all of the gaps surrounding bricks. With gaps somewhat filled, use a broom to sweep mortar around until all of them are evenly filled and piles of excess mortar are removed.

Next, apply a light water spray over brick insert areas to dampen mortar. Do not flood areas with a lot of heavy water applications, as this will only cause mortar to float away. Rely on a number of light mist applications to slowly penetrate mortar without disturbing its position. Mist on

18-44 Various brick insert designs are successfully installed by innovative concrete finishers.

and off every few minutes until mortar will no longer accept any new moisture.

Remove excess mortar from bricks with a small brush or putty knife while it is wet. Later, once mortar has had some time to cure, you can carefully use a small amount of water and a brush to clean bricks. Light applications with a wire brush may be needed for stubborn mortar accumulations.

POWER TROWEL FINISHING

Professional concrete finishers use gasoline-powered concrete finishing machines to finish off huge slabs. They are very worthwhile timesaving tools for large jobs encompassing hundreds of square feet of concrete. You might have seen them used for commercial building projects. Their need and usefulness for normal homeowner slabs is questionable, however.

Power trowels feature trowel-like finishing blades located below a powerful engine. Their speed is adjusted with a hand throttle. Machines are maneuvered by slightly twisting and moving the handle up or down. Proficient use requires practice.

Concrete must be well cured before machines are used because they are heavy and spin with such great force. Putting them on slabs that are still too wet will just cause a great deal of cream to be worked up, resulting in a sloppy finish. As a general rule, wait until concrete is hard enough for you to walk on before employing a power trowel.

Finally, machines must be moved around quickly enough to ensure a smooth finish and also avoid prolonged finishing in just one spot. Too much finishing activity tends to wipe out cream and bring up sand. Practice with these machines is a must. Although they are available at most rental yards, refrain from their use unless you have experience with them and are planning to pour a huge garage that will feature simply a plain, smooth trowel finish.

Index

Other Bestsellers of Related Interest

WHOLE HOUSE REMODELING GUIDE
S. Blackwell Duncan

This book features hundreds of remodeling, renovating, and redecorating options described and illustrated step-by-step! Focusing on interior modeling, the possibilities that exist for floors, windows, doors, walls, and ceilings are comprehensively explored. Complete detailed, illustrated instructions for projects are easy to follow. 448 pages, illustrated. **Book No. 3281, $19.95 paperback, $28.95 hardcover**

BE YOUR OWN ARCHITECT
Gene B. Williams

Cut the cost of home ownership by custom designing your own home. This book shows you how to turn your vague ideas of this ideal home into actual house plans. Williams makes you think like an architect—you'll quickly see the importance of considering such elements as traffic flow, storage space, heating, and cooling in your design. Information on title searches, insurance, materials and construction costs, and even tips on how to draw your plans correctly, has all been included. 288 pages, 300 illustrations. **Book No. 3336, $16.95 paperback, $26.95 hardcover**

THE COMPLETE BOOK OF LOCKS AND LOCKSMITHING—3rd Edition
C. A. Roper and Bill Phillips

In this profusely illustrated handbook, you find the knowledge you need to select, install, and maintain a wide variety of locks and security hardware. Whether you're a beginning locksmith or budget-conscious do-it-yourselfer . . . here is the clear, practical approach you need to understand your home or business security system. 448 pages, 620 illustrations. **Book No. 3522, $19.95 paperback, $28.95 hardcover**

FINISH CARPENTRY ILLUSTRATED
Elizabeth and Robert Williams

This guide provides detailed instruction on completing all of the visible refinements that give your work that polished look. You'll learn step-by-step how to install doors, windows, shelves, shutters, bathroom and kitchen cabinets, moldings, and more. And you'll apply finishing touches without the frustration that often goes with delicate work. 192 pages, 74 illustrations. **Book No. 3434, $12.95 paperback only**

HARDWOOD FLOORS—2nd Edition
Dan Ramsey

Hardwood floors have begun to reappear in new and remodeled homes because they are both beautiful and practical. If you yearn for the mellow look of wood, let Ramsey guide you through the selection of the proper tools and materials. He explains how to install or restore plank, tongue and groove, parquet, and block hardwood floors and then finish them to match your home's special decor. 192 pages, 222 illustrations, 4 full-color pages. **Book No. 3529, $14.95 paperback only**

HOME HEATING AND AIR CONDITIONING SYSTEMS
James L. Kittle

Spare yourself the aggravation of trying to locate a repairman when your furnace or air conditioning system breaks down—do your own professional-quality maintenance and repair! With the comprehensive instruction and guidance included here, you can install, repair, or replace just about any type of home heating or air conditioning system—oil, gas, hot water, or forced air heating systems, and central air systems, heat pumps, and more. 272 pages, 198 illustrations. **Book No. 3257, $15.95 paperback, $24.95 hardcover**

HOW TO PLAN, CONTRACT, AND BUILD YOUR OWN HOME—2nd Edition
Richard M. Scutella and Dave Heberle,
Illustrated by Jay Marcinowski

In this revised edition, soon-to-be homeowners will find the information they need to plan, contract, and build a home to their specifications. Covering the entire decision-making process, this guide outlines many of the important details you should consider before building a new home. New material includes helpful information on: laundry and utility room design, plumbing, specialty homes, garages, maintenance programs, and burglar-proofing. 424 pages, 300 illustrations. **Book No. 3584, $16.95 paperback only**

MAINTAINING AND REPAIRING VCRs —2nd Edition
Robert L. Goodman

". . . of immense use . . . all the necessary background for learning the art of troubleshooting popular brands," said **Electronics for You** about the first edition of this indispensable VCR handbook. Revised and enlarged, this illustrated guide provides complete professional guidance on troubleshooting and repairing VCRs from all the major manufacturers, including VHS and Betamax systems and color video camcorders. Includes tips on use of test equipment and servicing techniques plus case history problems and solutions. 352 pages, 427 illustrations. **Book No. 3103, $17.95 paperback, $27.95 hardcover**

ROOFING THE RIGHT WAY—2nd Edition
Steven Bolt

Why pay a contractor thousands of dollars to put a new roof on your home when you can do it yourself? Roofing isn't as difficult or as costly as you might think. Following the guidelines presented here, you can install a watertight roof that will add to the beauty and value of your home for years. You'll find in-depth details on every aspect of roofing—from choosing the proper tools and materials to step-by-step application techniques for nearly any type of roof. 240 pages, 277 illustrations. **Book No. 3387, $14.95 paperback, $24.95 hardcover**

ALL ABOUT CARPETS: A Consumer Guide
Glenn Revere

Comprehensive and concise, this illustrated consumer handbook explains carpet characteristics—types, weaves, backings—and examines carpet padding, dye techniques, and maintenance. Step-by-step instructions for carpet installation are also covered. A thorough and interesting guide to carpet purchasing and maintenance. 160 pages, 108 illustrations. **Book No. 2646, $9.95 paperback only**

DECKS AND PATIOS: Designing and Building Outdoor Living Spaces
Edward A. Baldwin

This handsome book will show you step by step how to take advantage of outdoor space. It's a comprehensive guide to designing and building decks and patios that fit the style of your home and the space available. You'll find coverage of a variety of decks, patios, walkways, and stairs. Baldwin helps you design your outdoor project, and then shows you how to accomplish every step from site preparation through finishing and preserving your work to ensure many years of enjoyment. 152 pages, 180 illustrations. **Book No. 3326, $16.95 paperback only**

HOME ELECTRICAL WIRING MADE EASY: Common Projects and Repairs
Robert Wood

"Written for people with no more electrical expertise than changing light bulbs, this book presents safe and easy procedures."

Popular Electronics

Following Wood's step-by-step instructions, you will become a master at installation, repair, and replacement of outlets, three- and four-way switches and timers, ceiling lights and fans, automatic garage door controls, thermostats, 220-volt appliance outlets, door bells, outdoor lighting, and more. 208 pages, 190 illustrations. **Book No. 3072, $16.95 paperback only**

DOORS, WINDOWS & SKYLIGHTS
—2nd Edition
Dan Ramsey

There's no end to the money you can save and the value and comfort you can add to your home with the practical ideas found in this well-illustrated guide. Ramsey gives step-by-step installation instructions for all kinds of attractive, economical interior and exterior doors, windows, and skylights. This edition features information on designing solariums and greenhouses, increasing energy efficiency with better insulating materials, and improving home security with locks and alarm systems. 240 pages, 282 illustrations. **Book No. 3248, $14.95 paperback only**

PRACTICAL HOUSEBUILDING:
For Practically Everyone
Frank Jackson
Illustrations by Spike Hendriksen

This idea-packed guide tackles the subject of housebuilding with humor and candor. You're walked through every area of knuckle-busting, do-it-yourself construction and you'll emerge chuckling but confident, with the same know-how as those who learned the hard way. Step-by-step instructions and over 250 detailed illustrations show you how to design and build your own home. 272 pages, illustrated. **Book No. 3808, $14.95 paperback only**

Prices Subject to Change Without Notice.

Look for These and Other TAB Books at Your Local Bookstore

To Order Call Toll Free 1-800-822-8158
(in PA, AK, and Canada call 717-794-2191)

or write to TAB Books, Blue Ridge Summit, PA 17294-0840.

Title	Product No.	Quantity	Price

☐ Check or money order made payable to TAB Books

Charge my ☐ VISA ☐ MasterCard ☐ American Express

Acct. No. _____ Exp. _____

Signature: _____

Name: _____

Address: _____

City: _____

State: _____ Zip: _____

Subtotal $ _____

Postage and Handling
($3.00 in U.S., $5.00 outside U.S.) $ _____

Add applicable state and local
sales tax $ _____

TOTAL $ _____

TAB Books catalog free with purchase; otherwise send $1.00 in check or money order and receive $1.00 credit on your next purchase.

Orders outside U.S. must pay with international money order in U.S. dollars.

TAB Guarantee: If for any reason you are not satisfied with the book(s) you order, simply return it (them) within 15 days and receive a full refund.   **BC**